Bioinformatics:
Applications in Life and Environmental Sciences

Bioinformatics:
Applications in Life and Environmental Sciences

Edited by

M.H. Fulekar

Department of Life Sciences
University of Mumbai, India

A C.I.P. Catalogue record for this book is available from the Library of Congress.

ISBN   978-1-4020-8879-7 (HB)
ISBN   978-1-4020-8880-3 (e-book)

Copublished by Springer,
P.O. Box 17, 3300 AA Dordrecht, The Netherlands
with Capital Publishing Company, New Delhi, India.

Sold and distributed in North, Central and South America by Springer,
233 Spring Street, New York 10013, USA.

In all other countries, except India, sold and distributed by Springer, Haberstrasse
7, D-69126 Heidelberg, Germany.

In India, sold and distributed by Capital Publishing Company,
7/28, Mahaveer Street, Ansari Road, Daryaganj, New Delhi, 110 002, India.

www.springer.com

Printed on acid-free paper

Printed in India.

Preface

Bioinformatics, a multidisciplinary subject, has established itself as a full-fledged discipline that provides scientific tools and newer insights in the knowledge discovery process in the front-line research areas and has become a growth engine in Biotechnology. It provides a better understanding of biomolecules, which is applicable at all levels of the living world starting from atomic and molecular levels to the higher complex of systems biology. The discovery of biomolecular sequences and their relation to the functioning of organisms have created a number of challenging problems for computer scientists, and led to emerging interdisciplinary field of Bioinformatics, Computational Biology, Genomics and Proteomics studies. Bioinformatics (Computational Biology) is a relatively new field that applies computer science and information technology to biology. In recent years, the discipline of Bioinformatics has allowed the biologist to make full use of the advances in computer sciences and computational statistics for advancing the biological data. Researchers in the Life Sciences generate, collect and analyze an increasing number of different types of scientific data, DNA, RNA and protein sequences, in situ and microarray gene expression including 3D protein structures and biological pathways.

This book aims at providing information on Bioinformatics at various levels to the post-graduate students and research scholars studying in the areas of Biotechnology/Environmental Sciences/Life Sciences and other applied and allied biosciences and technologies. The chapters included in the book cover from introductory to advanced aspects, including applications of various documented research work and specific case studies related to Bioinformatics. The topics covered under Bioinformatics in Life and Environmental Sciences include: Role of Computers in Bioinformatics, Comparative Genomics and Proteomics, Bioinformatics – Structural Biology Interface, Statistical Mining of Gene and Protein Databanks, Building Bioinformatics Database Systems, Bio-sequence signatures using Chaos Game Representation, data mining for Bioinformatics, Environmental clean up approach using Bioinformatics in Bioremediation and Nanotechnology in relation to Bioinformatics. This book will also be of immense value to the readers of different backgrounds such as engineers, scientists, consultants and policy makers for industry, government, academics and social and private organizations. Bioinformatics is an innovative field in the area of computer science and information technology for interpretation and compilation of data and gives the opening for advanced research and its applications in biological sciences.

The book has contributions from expert academicians, scientists from Universities, Institutes, IIT, BARC and personnel from research organizations and industries. This publication is an attempt to provide information on Bioinformatics and its application in life and environmental sciences.

I am grateful to Honourable Vijay Khole, the Vice Chancellor of University of Mumbai for having provided facilities and encouragement. It gives me immense pleasure to express deep gratitude to my colleagues and friends who have helped me directly or indirectly in completing this manuscript and also to my family in particular my wife Dr. (Mrs.) Kalpana and children Jaya, Jyoti and Vinay and brothers Dilip and Kishor for their constant support and encouragement. The technical assistance provided by my PhD students is also greatly acknowledged.

M.H. Fulekar
Professor and Head
Dept. of Life Sciences
University of Mumbai
Mumbai, India

Contributors

M.H. Fulekar
Professor & Head
Department of Life Sciences
Univ. of Mumbai, Mumbai-400 098
Email: mhfulekar@yahoo.com

B.B. Meshram
Professor
Computer Engineering Division
VJTI, Matunga
Mumbai-400 019
Email: bbmeshram@vjti.org.in

Rajani R. Joshi
Professor
Department of Mathematics
IIT Mumbai, Powai
Mumbai-400 076
Email: rrj@math.iitb.ac.in

M.V. Hosur
Senior Scientist
Solid State Physics Division (SSPD)
BARC, Trombay
Mumbai-400 085
Email: hosur@barc.gov.in

T.M. Bansod
Consultant, Information Security
MIEL e-security Pvt. Ltd.
Andheri (E), Mumbai
Email: tmbansod@hotmail.com

Achuthsankar S. Nair
Director, Centre for Bioinformatics
University of Kerala, Trivandrum
Email: sankar.achuth@gmail.com

Vrinda V. Nair
Centre for Bioinformatics
University of Kerela, Trivandrum

K.S. Arun
Centre for Bioinformatics
University of Kerela, Trivandrum

Krishna Kant
Department of Information Technology
Govt. of India, New Delhi

Alpana Dey
Department of Information Technology
Govt. of India, New Delhi

S.I. Ahson
Professor & Head
Department of Computer Science
Jamia Millia Islamia
New Delhi
Email: drsiahson@yahoo.com

T.V. Prasad
Professor & Head
Dept. of Computer Science &
Engineering
Lingaya's Institute of Management &
Technology
Faridabad-121 002, Haryana
Email: tvprasad2002@yahoo.com

Desh Deepak Singh
Indian Institute of Advanced Research
Koba, Gandhinagar
Gujarat-382 007
Email: ddsingh@iiar.res.in

S.J. Gupta
Professor & Head
Department of Physics
University of Mumbai
Mumbai-400 098
Email: gupta1947@rediffmail.com

Contents

1 Bioinformatics in Life and Environmental Sciences

M.H. Fulekar

INTRODUCTION

Bioinformatics involves close relation between biology and computers that influence each other and synergistically merging more than once. The variety of data from biology, mainly in the form of DNA, RNA, protein sequences is putting heavy demand in computer sciences and computational biology. It is demanding transformation of basic ethos of biological sciences.

The Bioinformaticians are those who specialize in use of computational tools and systems to answer problems in biology. They include computer scientists, mathematicians, statisticians, engineers and biologists who specialize in developing algorithms, theories and techniques for such tools and systems. Bioinformatics has also taken on a new glitter by entering the field of drug discovery in a big way. System biology promises great growth in modeling silver line using engineering approach with close relation with biology. Bioinformatic techniques have been developed to identify and analyze various components of cells such as gene and protein function, interactions and metabolic and regulatory pathways. The next decade will belong to understanding cellular mechanism and cellular manipulation using the integration of bioinformatics, wet lab, and cell simulation techniques. Bioinformatics has focus on cellular and molecular levels of biology and has a wide application in life sciences. Current research in bioinformatics can be classified into: (i) genomics—sequencing and comparative study of genomes to identify gene and genome functionality, (ii) proteomics—identification and characterization of protein related properties, (iii) cell visualization and simulation to study and model cell behaviour, and (iv) application to the development of drugs and anti-microbial agents. Bioinformatics also offers many interesting possibilities for bioremediation from environment protection point of view. The integration of biodegradation information, with the corresponding protein and genomic data, provides a suitable framework for studying the global properties of the bioremediation network. This discipline requires the integration of huge amount of data from various sources—

chemical structure and reactivity of organic compounds, sequence, structure and function of proteins (enzymes), comparative genomics, environment microbiology and so on. Bioinformatics has the following major branches: Genomics, Proteomics, Computer Aided Drug Designing, Biodatabase and Data Mining, Molecular Phylogenetics, Microarray informatics and System biology. Thus, bioinformatics has focused on cellular and molecular levels of biology and has a wide application in life sciences and environment protection.

The study and understanding of **cell** is prime concern for bioinformatics.

CELL COMPONENTS

The *cell* is the structural and functional unit of all living organisms, and is sometimes called the "building block of life". Some organisms, such as bacteria, are unicellular, consisting of a single cell. Other organisms, such as humans, are multicellular. (Humans have an estimated 100 trillion or 10^{14} cells; a typical cell size is 10 μm; a typical cell mass is one nanogram.) All living organisms are divided into five kingdoms: Monera, Protista, Fungi, Plantae and Animalia.

All cells fall into one of the two major classifications of prokaryotes and eukaryotes.

Prokaryotes	*Eukaryotes*
Prokaryotes are unicellular organisms, found in all environments. Prokaryotes are the largest group of organisms, mostly due to the vast array of bacteria which comprise the bulk of the prokaryote classification.	Eukaryotes are generally more advanced than prokaryotes. There are many unicellular organisms which are eukaryotic, but all cells in multicellular organisms are eukaryotic.
Characteristics: • No nuclear membrane (genetic material dispersed throughout cytoplasm) • No membrane-bound organelles • Simple internal structure • Most primitive type of cell (appeared about four billion years ago)	*Characteristics*: • Nuclear membrane surrounding genetic material • Numerous membrane-bound organelles • Complex internal structure • Appeared approximately one billion years ago
Examples: • *Staphylococcus* • *Escherichia coli* (E. coli) • *Streptococcus*	*Examples*: • *Paramecium* • *Dinoflagellates*

Chromosomes are the basic components of a cell. It is usually in the form of chromatin and composed of DNA which contains genetic information. The main component is *mitochondria* which is known as "power house" of the cell. This is a highly polymorphic organelle of eukaryotic cells that varies from short rod-like structures present in high number to long branched structures. It contains DNA and mitoribosomes and has double membrane system and the inner membrane may contain numerous folds (cristae). The inner fluid phase has most of the enzymes of the tricarboxylic acid cycle and some of the urea cycle. The inner membrane contains the components of the electron transport chain. Its major function is to regenerate ATP by oxidative phosphorylation. The Krebs cycle occurs within the central matrix of the mitochondrion and the cytochrome system occurs on the cristae which is bound on the large surface area of the sausage shaped mitochondrion. They are found in highly active cells like liver cells but not found in less active cells like red blood cells. Glycosylation and packaging of secreted proteins takes place in *Golgi apparatus*, which is an intracellular stack of membrane bounded vesicles. It is a net-like structure in the cytoplasm of animal cells (especially in those cells that produce secretions). Golgi apparatus is a cell organelle named after Camillo Golgi (1898), who first described it. The golgi apparatus receives newly synthesized molecules from the endoplasmic reticulum and stores them. It also attaches extra components to the molecule, such as adding carbohydrate to a protein (glycosylation). When the molecule is in demand, it is secreted in the form of a vesicle, which contains the finished product.

Nucleus is the major organelle of eukaryotic cells, in which the chromosomes are separated from the cytoplasm by the nuclear envelope. The nucleus has three parts: the nucleolus, the chromatin, and the nuclear envelope. A specialized, usually spherical mass of protoplasm encased in a double membrane, and found in most living eukaryotic cells, it directs their growth, metabolism, and reproduction, and functioning in the transmission of genic characters.

DNA (Deoxyribonucleic Acid)

The human body is estimated to contain 10 trillion cells and at some stage in its life cycle each contains a full complement of genes needed by the entire organism. Genes, composed of DNA in the nucleus, are clustered together in the chromosomes. In the chromosomes of all but the most primitive organism, DNA combines with protein. DNA, the molecular basis of heredity in higher organisms, is made up of a double helix held together by hydrogen bonds between purine and pyramidine bases i.e. between Adenine and Thymine and between Guanine and Cytosine.

The sequences of A, C, T and G could have a combination like:

TCCTGAT	AAGTCAG	TGTCTCCT	GAGTCTA	GCTTCTG	TCCATGC
TGATCAT	GTCCATG	TTCTAGT	CATGATA	GTTGATTC	TAGTCC
TGATTAG	CCTTGA	ATCTTCT	AGTTCT	GTCCAT	TATCCAT

Biotechnologists are able to isolate the cell, 'cut', 'open' the nucleus, and pull out the genome to read it using sequencing. These DNA sequencing is the complete blue print of life, including indication of what diseases the persons are susceptible to, and may be even predict your infidelity. Moreover it is the whole history book of life on the earth. The cell of our body has this information. The bioinformatic tools help to identify genome sequences and characterize for its significance in life.

RNA (Ribonucleic Acid)

RNA are similar to DNA. The major function of RNA is to copy information from DNA and bring it out of the nucleus to use wherever it is required to be used. The RNA in the cell has at least four different functions.

1. *Messenger RNA* (mRNA) is used to direct the synthesis of specific proteins.
2. *Transfer RNA* (tRNA) is used as an adapter molecule between the mRNA and the amino acids in the process of making the proteins.
3. *Ribosomal RNA* (rRNA) is a structural component of a large complex of proteins and RNA known as the ribosome. The ribosome is responsible for binding to the mRNA and directing the synthesis of proteins.
4. The fourth class of RNA is a catch-all class. There are small, stable RNAs whose functions remain a mystery. Some small, stable RNAs have been shown to be involved in regulating expression of specific regions of the DNA. Other small, stable RNAs have been shown to be part of large complexes that play a specific role in the cell. In general, RNA is used to convey information from the DNA into proteins.

Like DNA, RNA contains four kinds of molecules—A, G, C and U—the last one replacing Thymine in DNA. An RNA sequence may run like this:

UCCUGAU	AAGUCAG	UGUCUCCU	GAGUCUA	GCUUCUG
UCCAUGC	UGAUCAU	GUCCAUG	UUCUAGU	CAUGAUA
GUUGAUUC	UAGUGUCC	UGAUUAG	CCUUGA	AUCUUGA
AUCUUCU	AGUUCU	GUCCAU	UAUCCAU	

RNA is of different kinds. Biologists have to text a file of their sequences and make use in Life Sciences.

PROTEINS AND AMINO ACIDS

The most distinguishing features of reactions that occur in a living cell are the participation of enzymes as biological catalysts. Almost all enzymes are proteins and have globular structure and many of them carry out their catalytic function by relying solely on their protein structure. Many others require non-protein components called cofactors, which may be metal ions or organic molecules referred to as 'Coenzymes'.

Proteins are made up of amino acids which are 20 in number. These are Alanine, Arginine, Asparginine, Aspartic acid, Cysteine, Glutamic acid, Glutamine, Glysine, Valine, Serine, Leucine, Isolucine, Methionine, Phenylalanine, Tyrosine, Tryptophane, Proline, Ornithine, Taurine, and Threonine. Proteins carry both positive and negative charges on their surface, largely due to the side chains of acidic and basic amino acids. Positive charges are contributed by Histidine, Lysine and Arginine and to a lesser extent, N-terminal amino acids; negative charges are due to aspartic and glutamic acids, C-terminal carboxyl groups and to a lesser extent cysteine residues. The net charge on protein depends on the relative numbers of positive and negative charged groups. This varies with pH. The pH where a protein has an equal number of positive and negative charged groups is termed as iso-electric point (pI). Most proteins have a net negative charge while below it their charge is positive.

A protein sequence will look like:

CFPUGEGHILDCLKSTFEWCUWECFPWRDTCEDUSTTWEGHILD NDTEGHTWUWWESPUSTPPUGWRDCCLKSWCUWMFCQEDTW RWEGHILKMFPUSTWYZEGNDTWRDCFPUQEGHILDCLKSTMF EWCUWESTHCFPWRDT

The gene regions of the DNA in the nucleus of the cell is copied (transcribed) into the RNA and RNA travels to protein production sites. DNA-RNA-protein is the central dogma of molecular biology. Each DNA is crunching out thousands of RNAs which in turn cause thousands of proteins to be produced. That makes the characteristics of the persons.

GENOMICS

Life sciences have witnessed many important events over the last couple of decades, not least of which is the advent of genomics. Lockhart et. al. have noted that:

...the massive increase in the amount of DNA sequence information and the development of technologies to exploit its use...(have prompted) new types of experiments...observations, analysis and discoveries...on an unprecedented scale...Unfortunately, the billions of bases of DNA sequence do not tell us what all the genes do, how cells work, how cells form organisms, what goes wrong in disease, how we age or how to develop a drug...The purpose of genomics is to understand biology not simply to identify its component parts (Lockhart and Winzeler, 2000).

Genomics is the study of the complete set of genetic information—all the DNA in an organism. This is known as its genome. Genome range is size: the smallest known bacterial genome contains about 600,000 base pairs and the human genome has some three billion. Typically, genes are segments of DNA that contain instructions on how to make the proteins that code for structural and catalytic functions. Combinations of genes, often interacting with environmental factors, ultimately determine the physical characteristics of an organism (NABIR, 2003). At the DOE joint genome institute (USA) and other sequencing centre, procedures have been developed which enable rapid sequencing of microorganisms. Microbial genomes are first broken into shorter pieces. Each short piece is used as a template to generate a set of fragments that differ in length from each other by a single base. The last base is labelled with a fluorescent dye specific to each of the four base types. The fragments in a set are separated by gel electrophoresis. The final base at the end of each fragment is identified using laser-induced fluorescence which discriminates among the different labelled bases. This process recreates the original sequence of bases (A, T, G and C) for each short piece generated in the first step. Automated sequences analyze the resulting electrophorograms and the output is a four-colour chromogram showing peaks that represent each of the four DNA bases. After the bases are "read", computers are used to assemble the short sequences into long continuous stretches that are analyzed for errors, gene-coding regions and then characteristics. To generate a high quality sequence, additional sequencing is needed to close gaps, and allow for only a single error in every 10,000 bases. By the end of the process, the entire genome would have been sequenced, putting that defined features of biological importance must be identified and annotated. When the newly identified gene has a close relative already in a DNA database, gene finding is relatively straight forward. The genes tend to be single, uninterrupted open reading frames (ORFs) that can be translated and compared with the database. Scientists in the new discipline of bioinformatics are developing and applying computational tools and algorithms to help identify the functions of these previously unidentified genes. An accurate description of genes in microbial genomes is essential in describing metabolic pathways and often aspects of whole organism's functions (NABIR, 2003).

Four interlinked areas of interest have emerged from these efforts:

- *Genome mapping and sequencing* to locate genes on the chromosomes (Boguski and Schuler, 1995) and determining the exact order of the nucleotides that make up the DNA of the chromosomes (International Human Genome Sequencing Consortium, 2001).
- *Structural genomics to provide* a model structure for all of the tractable macromolecules that are encoded by complete genomes (Bremer, 2001).
- *Functional genomics* to flesh out the relationships between genes and phenotypes (White, 2001).
- *Comparative genomics* to identify various functional sequences (Bofelli, Nobrega and Rubin, 2004).

Genomics is the complete analysis of DNA sequences that contain the codes for hereditary behaviours and proteomics. The analysis of complete sets of proteins have been the main beneficiaries of bioinformatics.

PROTEOMICS

Proteomic is the study of the many and diverse properties of proteins in parallel manner with the aim of providing detailed descriptions of the structure, function and control of biological systems in health and disease (Patterson and Aebersold, 2003).

Proteins are not involved in the majority of all biological processes but collectively contribute significantly to the understanding of biological system. It comprises protein sequence data, the set of protein sequences and finding their similarity, pairwise and multiple sequence alignment, determining primary, secondary and tertiary structures of molecule, protein folding problem, finding the chemically active part of protein, interaction with another protein, protein-protein interaction, identification of cell component to which protein belongs, proteins sub-cellular localization and protein sorting problems. The sequence alignment technique is widely applied in both genomics and proteomics. Since it is very difficult and expensive to evaluate structures by methods like x-ray diffraction or NMR spectroscopy, there is a big need for unfailing prediction of 3D structure of protein sequence data. Prediction of protein structure from sequences is one of the most challenging tasks in today's computational Biology which is made possible by bioinformatics. Phylogenetic trees are building up with the information gained from the comparison of the amino acid sequences of a protein like Cytochrome C sampled from different species. The phylogenetic tree can be created by prediction of protein structures. Proteins of one class often show a few amino acids that always occur at the same position in the amino acid sequences. By looking for "Patterns" you will be able to gain information about the activity of a protein of which only the gene (DNA) is known. Evaluation of such patterns yields information about the architecture of proteins. The sequence comparison is a very powerful tool in molecular biology, genetics and proteomics chemistry.

APPLICATION IN ENVIRONMENT

The application of molecular biology-based techniques is being increasingly used and has provided useful information for improving the strategies and assessing the impact of treatment technology on ecosystems. Recent developments in molecular biology techniques also provide rapid, sensitive and accurate methods by analyzing bacteria and their catabolic genes in the environment. The sustainable development requires the promotion of environmental management and a constant search for new technologies to treat vast quantities of wastes generated by increasing anthropogenic activities. Biotreatment, the processing of wastes using living organisms, is an environment friendly, relatively simple and cost-effective alternative to physico-chemical clean-up options. Confined environments, such as bioreactors, have been engineered to overcome the physical, chemical and biological-limiting factors of biotreatment processes in highly controlled systems. The great versatility in the design of confined environments allows the treatment of a wide range of wastes under optimized conditions. To perform a correct assessment, it is necessary to consider various microorganisms having a variety of genomes and expressed transcripts and proteins. A great number of analyses are often required. Using traditional genomic techniques, such assessments are limited and time-consuming. However, several high-throughput techniques originally developed for medical studies can be applied to assess biotreatment in confined environments.

Bioremediation also takes place in natural environment. However, the natural biodegradation of the xenobiotics present in the environment is a slow process. Natural attenuation is one of several cost-saving options for the treatment of polluted environment, in which microorganisms contribute to pollutant degradation. For risk assessments and endpoint forecasting, natural attenuation sites should be carefully monitored (monitored natural attenuation). When site assessments require rapid removal of pollutants, bioremediation, categorized into biostimulation (introduction of nutrients and chemicals to stimulate indigenous microorganisms) and bioaugmentation (inoculation with exogenous microorganisms), can be applied. In such a case, special attention should be paid to its influences on indigenous biota and the dispersal and outbreak of inoculated organisms. Recent advances in microbial ecology have provided molecular technologies, e.g., detection of degradative genes, community fingerprinting and metagenomics, which are applicable to the analysis and monitoring of indigenous and inoculated microorganisms in polluted sites. Scientists have started to use some of these technologies for the assessment of natural attenuation and bioremediation in order to increase their effectiveness and reliability.

The study of the fate of persistent organic chemicals in the environment has revealed a large reservoir of enzymatic reactions. Novel catalysts can be obtained from metagenomic libraries and DNA-sequence based approaches. The increasing capability in adapting the catalysts to specific reactions and process requirements by rational and random mutagenesis broadens the scope for environmental clean up employing biodegradation in the field. However, these catalysts need to be exploited in whole cell bioconversions or in fermentations, calling for system-wide approaches to understanding strain physiology and metabolism and rational approaches to the engineering of whole cells as they are increasingly put forward in the area of environment biotechnology.

The efficient utilization of aromatic compounds by bacteria are the enzymes which are responsible for their degradation and the regulatory elements that control the expression of the catabolic operons to ensure the more efficient output depending on the presence/absence of the aromatic compounds or alternative environmental signals. Transcriptional regulation seems to be the more common and/or most studied mechanism of expression of catabolic clusters, although post-transcriptional control also plays an important role. Transcription is dependent on specific regulators that channel the information between specific signals and the target gene(s). A more complex network of signals connects the metabolic and the energetic status of the cell to the expression of particular catabolic clusters, overimposing the specific transcriptional regulatory control. The regulatory networks that control the operons involved in the catabolism of aromatic compounds are endowed with an extraordinary degree of plasticity and adaptability. The regulatory networks elucidate the way for a better understanding of the regulatory intricacies that control microbial biodegradation of aromatic compounds.

Microarray technology has the unparalleled potential to simultaneously determine the dynamics and/or activities of most, if not all, of the microbial populations in complex environments such as soils and sediments. Researchers have developed several types of arrays that characterize the microbial populations in these samples based on their phylogenetic relatedness or functional genomic content. Several recent studies have used these microarrays to investigate ecological issues; however, most have only analyzed a limited number of samples with relatively few experiments utilizing the full high-throughput potential of microarray analysis. Microarrays have the unprecedented potential to achieve this objective as specific, sensitive, quantitative, and high-throughput tools for microbial detection, identification, and characterization in natural environments. Due to rapid advances in bioinformatics, microarrays can now be produced that contain thousands to hundreds of thousands of probes. Microarrays have been primarily developed and used for gene expression profiling of pure cultures of individual organisms, but major advances have recently been made in their application to environmental samples. However, the analysis of environmental samples presents several challenges not encountered during the analysis of pure cultures. Like most other techniques, microarrays currently detect only the dominant populations in many environmental samples. In addition, some environments contain low levels of biomass, making it difficult to obtain enough material for use in microarray analysis without first amplifying the nucleic acids. Such techniques, even if applied with utmost care, may introduce biases into the analyses, but perhaps the greatest challenge to the analysis of environmental samples using microarrays is the vast number of unknown DNA sequences in these samples. The importance of an organism, which may be dominant and critical to the ecosystem under study, can be completely overlooked if the organism does not have a corresponding probe on the array. Probes designed to be specific to known sequences can also cross-hybridize to similar, unknown sequences from related or unrelated genes, resulting in either an underestimated signal due to weaker binding of a slightly divergent sequence or a completely misleading signal due to binding of a different gene. Furthermore, it is often a challenge to analyze microarray results from environmental samples due to the massive amounts of data generated and a lack of standardized controls and data analysis procedures. Despite these challenges, several types of microarrays have been successfully applied to microbial ecology research.

The numerous techniques can be combined with microarray analysis not only to validate results but also to produce powerful synergy for investigating microbial interactions and processes. Microarray technologies and the integration of isotopes produce one of the potentially most powerful combined approaches for microbial ecology research. Microarray analysis of DNA or RNA labeled with isotopes can differentiate between active and inactive organisms in a sample and/or identify those organisms that metabolize a labeled substrate. These isotopes can be either radioisotopes such as 14C or stable isotopes such as 13C. Scientists used 14C-labeled bicarbonate and a POA to study ammonia-oxidizing bacterial communities in two samples of nitrifying activated sludge. Scanning for radioactivity in the rRNA hybridized to the POA enabled detection of populations that consumed the [14C] bicarbonate. The approach detected populations that composed less than 5–10% of the community. This technique could potentially be applied to 13C-labeled materials also, but it may be more difficult to obtain enough 13C-labeled DNA for microarray analyses since the labeled DNA would have to be separated from non-labeled DNA before microarray analysis unless the array could be directly scanned for C.

The recent bioinformatics in bioremediation studies have used microbial environment in environmental processes including nitrogen fixation, nitrification, denitrification and sulphate reduction in fresh water and marine systems; degradation of organic contaminants including polychlorinated biphenyls (PCBs) and polycyclic aromatic hydrocarbons (PAHs), in soil and sediments and methane oxidizing capacity and diversity in landfill-simulating soil. However, many of these applications were conducted primarily as proofs of concept and did not analyze enough samples or treatments to enable biological meaningful conclusions to be formed.

BIOINFORMATIC DATABASES

The rise of Genomics and Bioinformatics has had another consequence—the increasing dependence of biology on results available only in electronic form. Most of the useful genomic data, notably genetic maps, physical maps as well as DNA and protein sequences, are available only on the World–Wide–Web. Not only are these data unsuited because of their very bulk to print media, they are of very little use in print because that kind of information can only be truly assimilated, used and appreciated with the aid of computers and software. This trend is rapidly being extended to nonsequence data such as mutant phenotypes, gene expression patterns and gene interactions, whose complexity defies simple description. In all such descriptions, there are atleast as many data points as there are genes in an organism, meaning that we can look forward to data sets comprising literally millions of data points. Of necessity, results will only be summarized in print; the real data will reside as binary strings on electronic media. As a result, databases of genomic information for a variety of organisms have been organized (i.e. *Mycoplasma genitalium, Mycoplasma pneumoniae, Methanococcus jannaschii, Haemophilus influenzae* Rd, *Cynobacteria, Bacillus subtilis,* Mycobacteria, yeast, worm, *Drosophila, Arabidopsis,* maize, mouse and human).

Some Bioinfromatics Databases and their websites

Name of the database	Link of the database
DNA Data Bank of Japan (DDBJ)	http://www.ddbj.nig.ac.jp
EMBL Nucleotide Sequence Database	http://www.ebi.ac.uk/embl.html
GenBank	http://www.ncbi.nlm.nih.gov/
Genome Sequence Database (GSDB)	www.ncgr.org/gsdb
STACK	http://ww2.sanbi.ac.za/Dbases.html
TIGR Gene Indices	http://www.tigr.org/tdb/index.html
UniGene	http://www.ncbi.nlm.nih.gov/UniGene/
Clusters of Orthologous Groups (COG)	http://www.ncbi.nlm.nih.gov/COG/
ASDB	http://cbcg.nersc.gov/asdb
Axeldb	http://www.dkfz-heidelberg.de/abt0135/axeldb.htm
BodyMap	http://bodymap.ims.u-tokyo.ac.jp/
EpoDB	www.cbil.upenn.edu/EpoDB
FlyView	http://pbio07.uni-muenster.de/
Gene Expression Database (GXD)	http://www.informatics.jax.org/searches/gxdindex_form.shtml
Interferon Stimulated Gene Database	http://www.lerner.ccf.org/labs/williams/xchip-html.cgi
MethDB	http://www.methdb.de
Mouse Atlas and Gene Expression Database	http://genex.hgu.mrc.ac.uk

CONCLUSION

The availability of genomic and proteomic data and improved bioinformatics and biochemical tools have raised the expectation of humanity to be able to control the genetics by manipulating the existing microbes. Bioinformatics analysis will facilitate and quicken the analysis of systemic level of behaviour of cellular processes, and to understanding the cellular processes in order to treat and control microbial cells as factories. For the last decade, bioinformatic techniques have been developed to identify and analyze various components of cells such as gene and protein function, interactions and metabolic and regulatory pathways. The next decade will belong to understanding cellular mechanism and cellular manipulation using the integration of bioinformatics. The major impact of bioinformatics has been in automating the microbial genome sequencing, the development of integrated databases over the internet, and analysis of genomes to understand gene and genome function. Most of the bioinformatic techniques are critically dependent upon the knowledge derived from wet laboratories and the available computational algorithms and tools.

2 Role of Computers in Bioinformatics

Tularam M. Bansod

INTRODUCTION

Bioinformatics is a new and rapidly evolving discipline that has emerged from the fields of experimental molecular biology and biochemistry, and from the artificial intelligence, database, and algorithms disciplines of computer science (http://www.cs.wright.edu/cse/research/facilities-room.phtml?room=307). There are two possibilities to increase research in Bioinformatics field, one way is to teach computer science to biologists, biotechnologists and the other way is to teach biology to computer scientists. I think both the departments are doing these things in their capacities. It is easy to teach new computer technologies to biologists.

If we had taken survey of computer usage in the Life Science departments or Biology departments at the university level, we will find that people are using computers for administrative work, typing some research paper and browsing internet. Life science departments lack following things to utilize computers or computer networks:

1. Lack of requirement analysis of respective department.
2. Proper storage space for displaying research papers on the university site or department sites is missing or partial.
3. Lack of indigenous protein servers or genome server or powerful database servers.
4. Lack of co-ordination of Computer Science department and Life Science department in the university.
5. Lack of course-ware design in the computer technology by the computer faculty for the biologist.
6. Course-ware design by the biologist in the biology field for computer scientist also missing.
7. Lack of focussed research.
8. Lack of creative human resource.

When life science department try to find maximum usage of computer network, they should use following steps:

1. Calculate bandwidth requirement of department.
2. Calculate storage requirement of department.
3. Design computer network from consultants.
4. Check the load on the web servers regularly.
5. Information security policies should be designed.
6. Procure the building blocks of computer networks for bioinformatics lab.
7. Calculate assets value of your research work and computer network resources.
8. Regularly do risk analysis and information audits to justify the expenditure on overall computing resources.
9. Train the human resource in latest computer as compulsory component of lab budget.

Many foreign universities have started undergraduate courses in bioinformatics to create human resource in this upcoming field. We are still struggling to get a person to teach bioinformatics to the postgraduate level. In the syllabus of bioinformatics of Colman College, USA (Table 1), we noticed that around 40% syllabus is based on computer networking (http://www.coleman.edu/academics/undergraduate/bioinformatics.php/).

Table 1: Syllabus of Bio-Informatics Technicians (BIT) course in Colman College, USA

Course	Title	Units
COM113	Intro to PCs and Networks	4
COM107	Intro to Programing	8
COM259	UNIX Fundamentals	8
COM287	Internet Programing I	4
BIT100	General Biology	4
COM115	Client Services & Support I	4
MAT162	Algebra I	4
BIT110	Cellular and Molecular Biology	4
MAT250	Statistics for Bioinformatics	4
NET130	TCP/IP Fundamentals	4
BIT120	Microbiology and Immunology	4
NET250	Networking Concepts	4
BIT200	MySQL & Oracle (DB Admin)	4
BIT240	Bioinformatics	4
BIT210	Biotechnology	4
BIT260	Sequence and Structure DB	4
BIT220	BioPerl	4
BIT250	Introduction to BioJava	4
BIT230	LAMP	4

In this paper we will study computer networking and its applications for bioinformatics.

ESSENTIAL COMPUTER NETWORKING FOR BIOLOGIST

What is computer networking? Connection between active and passive components of computing is called networking. What is network of network? Internet is called network of network. Internet has many different types of networks such as Ethernet, SONET, and ATM (Asynchronous Transfer Mode). Following are components of computer networking:

* NIC (Network Interface Card)
* Hub
* Switches—layers 3 and 4
* Router—Cisco, Juniper, Dax
* Computer—P-III, P-IV, Servers
* Cables—Guided/Unguided

When we attach hub or switch with computer nodes with the help of copper or optical cable, we find active and passive components in the network. Copper cable and optical cables are passive components. Wireless communication is based on unguided signaling. Computer nodes and hub switches are called active components. Networking building blocks start with NIC card. This is called Network Interface Card. Now-a-days this card is inbuilt on the mother board. This NIC is available in different speeds viz., 10 mbps, 100 mbps, 1 gbps. This indicate maximum speed of network processing on your machine. There are different manufacturers of network cards e.g., D-link, SMC, Intel, Dax, Realtek.

Hub is the device used to connect computers in the LAN (Local Area Network). It is available in 8 ports, 16 ports and 24 ports. It is running in the physical layer. We can attach 8, 16, 24 computers to the Hubs. Hub is used to divide the speed of network in the number of computers. If we attach eight computers to the Hub and if NIC is of 10 Mbps then every computer is able to send packet by 10/8 mbps speed. There is no IC (Integrate Circuit) to process table of physical addresses. There is only one collision domain in the Hub. In the switch there are multiple collision domains. Switch looks like Hub and this also comes in the 8, 16, 24 ports. Switch has IC to process packets and maintain physical addresses of computers. Generally physical address is 48 bit, written in the hexadecimal bit.

Router is many times a very important device to forward packets to the dissimilar networks. Router works on network layer. We can implement security feature also in the router which will help to stop malicious codes at the periphery of the network. We can use firewall and DMZ (De-Military Zone) to protect bio-informatics network also.

Where computer networking affecting Bio-informatics Lab?

- Bandwidth Management
- Security Management
- Database Management

Bandwidth Management

- Dialup lines – Under-utilized
- 56 kbps/4 kbps
- Lease line
- Virtual Private Network
- Triband (Broadband launched by MTNL)

Security Management

- Physical security—Access control
- Functional security—Process security
- Application security—Bioedit, Rasmol, Browser security settings, SSL

Database Management

- Physical database integrity
- Logical database integrity
- Access control
- Confidentiality
- Availability

What are the research themes in Bioinformatics?

Novel computational techniques for the analysis and modeling of complex biological systems involves, pattern recognition, evolutionary computation, combinatorial and discrete analyses, statistical and probabilistic methods, molecular evolution, population genetics, protein structure, function, and assembly, and even forensic DNA analysis.

Peer-to-peer (P2P) networks have also been initiated to attract attention from those that deal with large datasets such as bioinformatics. P2P networks can be used to run large programs designed to carry out tests to identify drug candidates. The first such program was started in 2001 at the Centre for Computational Drug Discovery at Oxford University in cooperation with the National Foundation for Cancer Research. There are now several similar programs running under the auspices of the United Devices Cancer Research Project. On a smaller scale, a self-administered program for computational biologists to run and compare various bioinformatics software is available from Chinook. Tranche is an open-source set of software tools for setting up and administrating a decentralized network. It was developed to solve the bioinformatics data sharing problem in a secure and scalable fashion (http://

en.wikipedia.org/wiki/Peertopeer#Application_of_P2P_Network_outside_
Computer_Science).

Grid computing is a service-oriented architectural approach that uses open
standards to enable distributed computing over the Internet, a private network
or both. Grid computing helps to manage access to huge data, a crucial
capability for making rapid progress in life sciences research.

REFERENCES

Forouzan, B., TCP/IP Protocol. Tata McGraw Hill Publication, 2004.
Pfleeger, Charles, Security in Computing. Pearson Publication, 2004.
http://www.cs.wright.edu/cse/research/facilities-room.phtml?room=307
http://www.coleman.edu/academics/undergraduate/bioinformatics.php/
http://en.wikipedia.org/wiki/Peertopeer#Application_of_P2P_Network_outside_
    Computer_Science)

3 Comparative Genomics and Proteomics

M.V. Hosur

GENOMICS

Over the last decade, research in biology has undergone a qualitative transformation. Genome sequences, the bounded sets of information that guide biological development and function, lie at the heart of this revolution. The earlier reductionist approach to simplify complex biological systems has given way to an integrated approach dealing with the system as a whole, thanks to advances in various technologies. It is now possible to sequence, within a short time, entire genomes, which are the blue-prints, at the molecular level, for the complex workings of organisms. Genomes of more than 350 life-forms, including several mammals, have been sequenced, and many more are in the pipeline. The genomic sequence data is only the beginning, and to realize the full benefit of this data, the data has to be very skillfully processed and analysed. Comparative genomics is the study of relationships between the genomes. The broadly available genome sequences of human and a select set of additional organisms represent foundational information for biology and biomedicine. Embedded within this as-yet poorly understood code are the genetic instructions for the entire repertoire of cellular components, knowledge of which is needed to unravel the complexities of biological systems. Elucidating the structure of genomes and identifying the function of the myriad encoded elements will allow connections to be made between genomics and biology and will, in turn, accelerate the exploration of all realms of the biological sciences. Interwoven advances in genetics, comparative genomics, high throughput biochemistry and bioinformatics are providing biologists with a markedly improved repertoire of research tools that will allow the functioning of organisms in health and disease to be analysed and comprehended at an unprecedented level of molecular detail. In short, genomics has become a central and cohesive discipline of biomedical research.

By choosing appropriate genomes, the technique of comparative genomics has been applied to the following problems: 1. identify positions of genes in a genome, 2. annotate functions to genes through comparisons across species, 3. quantify evolutionary relationships between species, 4. identify targets for drug-development against diseases, and 5. probe the theory 'out of Africa' for human genesis.

PROTEOMICS

The term proteome was first coined to describe the set of proteins encoded by the genome. Proteome analysis is the direct measurement of proteins in terms of their presence and relative abundance. The overall aim of proteomics is the characterization of the complete network of cell regulation. Analysis is required to determine which proteins have been conditionally expressed, by how much, and what post-translational modifications have occurred.

The study of the proteome, called proteomics, now is almost everything 'post-genomic'. One can think of different types of proteomics depending on the technique used.

Mass Spectrometry-based Proteomics

Initial proteomics efforts relied on protein separation by two-dimensional gel electrophoresis, with subsequent mass spectrometric identification of protein spots. An inherent limitation of this approach is the depth of coverage, which is necessarily constrained to the most abundant proteins in the sample. Recently, the technology of Isotope Coated Affinity Tags (ICAT) has been developed that does away with electrophoresis thereby greatly improving sensitivity. This technology can easily yield information about how the concentration of proteins varies over the course of a cell's lifetime. The rapid developments in mass spectrometry have shifted the balance to direct mass spectrometric analysis, and further developments will increase sensitivity, robustness and data handling. The development of statistically sound methods for assignment of protein identity from incomplete mass spectral data (software SEQEST) will be critical for automated deposition into databases, which is currently a painstaking manual and error-prone process (Fig. 1).

Array-based Proteomics

A number of established and emergent proteome-wide platforms complement mass spectrometric methods. The forerunner amongst these efforts is the systematic two-hybrid screen developed by scientists. Unlike direct biochemical methods that are constrained by protein abundance, two-hybrid methods can often detect weak interactions between low-abundance proteins, albeit at the expense of false positives. Lessons learned from analysis of DNA microarray data, including clustering, compendium and pattern-matching approaches, should be transportable to proteomic analysis, and it is encouraging that the European Bioinformatics Institute and the Human

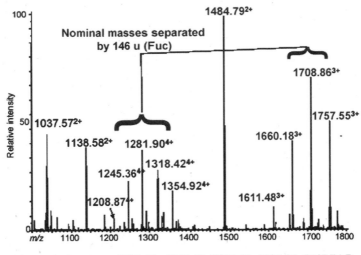

Peptide nominal masses: **4831.43; 4977.49; 5123.59; 5269.64; 5415.74 Da**

Fig. 1: Mass spectrum at 75 min from LC-MS of Amylase digest.

Proteome Organisation (HUPO) have together started an initiative on the exchange of protein–protein interaction and other proteomic data (*see* http://psidev.sourceforge.net/).

Structural Proteomics

Beyond a description of protein primary structure, abundance and activities, the goal of proteomics is to systematically understand the structural basis for protein interactions and function. A full description of cell behaviour necessitates structural information at the level not only of all single proteins, but of all salient protein complexes and the organization of such complexes at a cellular scale. This all-encompassing structural endeavour spans several orders of magnitude in measurement scale and requires a battery of structural techniques, from X-ray crystallography (Fig. 2) and nuclear magnetic resonance (NMR) at the protein level, to electron microscopy of mega-complexes and electron tomography for high-resolution visualization of the entire cellular milieu. NMR and *in silico* docking will be necessary to build in dynamics of protein interactions, much of which may be controlled through largely unstructured regions. The recurrent proteomic theme of throughput and sensitivity runs through each of these structural methods.

In the earlier days, crystal structure determinations were too slow to keep pace with sequence determinations. However, recently the following advances in technology have been achieved. Production and purification of recombinant proteins has been automated to a very large extent. Robots have been developed to rapidly set-up thousands of experiments to crystallize proteins.

The amount of reagents and protein used in these crystallization trials is reduced by as much as fifty times compared to manual trials. Very intense and highly collimated beams of x-rays are produced on synchrotrons enabling diffraction data collection from micron size crystals. Protein crystallography (Fig. 2) beamlines on these synchrotron sources incorporate automation at every step in the process of diffraction data collection. A complete diffraction data set can now be measured within minutes without needing any manual intervention. Some of the 'high-throughput' protein crystallography beamlines allow for remote operation, even making travel to the synchrotron unnecessary! Tunability of x-ray wavelengths at synchrotron sources has enabled routine solution of the 'Phase Problem' by exploiting the physical phenomenon of anomalous scattering. These developments have substantially increased the speed with which three dimensional structures of proteins can be determined. In many favourable cases, the three dimensional structure of a protein may be established much faster than conventional characterization of the protein through biochemical and genetic methods. Then a comparison of the three dimensional structure of the new protein with structures of other proteins with known functions may offer a faster route to annotate function to proteins.

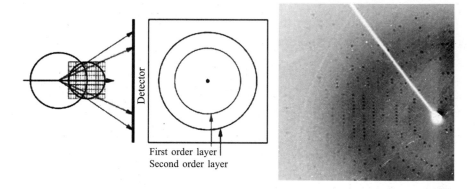

Fig. 2: Single crystal X-ray Diffraction

Proteomics is set to have a profound impact on clinical diagnosis and drug discovery. The selection of particular proteins for detailed structure analysis is dictated by results of comparative proteomics. Very often proteomics provides key input for selection of targets for drug-development. In the case of Plasmodium falciparum, comparative proteomics has identified hundreds of plasmodium specific proteins, which may provide targets for development of drugs and vaccines against malaria.

The technique of comparative proteomics has also contributed towards establishing protein and gene networking in cellular functioning, screening of lead molecules for toxicity and other effects.

Table 1: Comparative summary of the protein lists for each stage

Protein count	Sporozoites	Merozoites	Trophazoites	Gametocytes
152	×	×	×	×
197	–	×	×	×
53	×	–	×	×
28	×	×	–	×
36	×	×	×	–
148	–	–	×	×
73	–	×	–	×
120	×	–	–	×
84	–	×	×	–
80	×	–	×	–
65	×	×	–	–
376	–	–	–	×
286	–	–	×	–
204	–	×	–	–
513	×	–	–	–
2,415	1,049	839	1,036	1,147

Whole-cell protein lysates were obtained from, on average, 17×10^6 sporozoites, 4.5×10^9 traphozoites, 2.75×10^9 merozates, and 6.5×10^9 gametocytes.

METHODS OF COMPARISON

Comparison of genome sequences from evolutionarily diverse species has emerged as a powerful tool for identifying functionally important genomic elements. Assignment of function de-novo to a new protein is a very tedious process, and sequence comparisons may provide a short cut when sequence similarity is rather high. Structural comparisons, however, have a higher success rate while predicting protein function, because structural variations are slower compared to sequence variations. The key step in such comparative studies is the correct alignment of sequences being compared. The sheer amount of information contained in modern genomes and proteomes (several gigabytes in the case of humans) necessitates that the methods of *comparative genomics* and *comparative proteomics* be highly computational in nature. Parallel developments in the powers of computers has enabled researchers to develop softwares for the comparison of various types of genomic and proteomic data. Comparative methods involve two steps: (a) alignment of residues from the two sequences, and (b) scoring different alignments based on a chosen substitution matrix, M(I, J). One of the most dependable alignment methods is that developed by Needleman and Wunsch to compare protein sequences. This method for alignment of any two given sequences, A and B, consists of three steps: creation of an alignment matrix, M, in which the element M(i, j) represents the probability that residue at position i in sequence A is substituted with residue j in sequence B, modification of matrix M iteratively, using concepts of dynamic programing, in such a way that the modified value of M (i, j) encodes the best alignment path from the start to the current residue, and finally tracing a path in this matrix M to arrive at the complete alignment. Two types of substitution matrices, M(i, j), are currently in use: PAMn matrix and BLOSUMn matrix. PAM1 matrix estimates what rate of substitution would be expected if 1% of the amino acids had changed. The PAM1 matrix is used as the basis for calculating other matrices by assuming that repeated mutations would follow the same pattern as those in the PAM1 matrix, and multiple substitutions can occur at the same site. Using this logic, Dayhoff derived matrices as high as PAM250. Different PAM matrices are derived from the multiplication of the PAM1 matrix. A PAM250 is equivalent to one unchanged amino acid out of five. Sequence changes over long evolutionary time scales are not well approximated by compounding small changes that occur over short time scales. The BLOSUM62 matrix is calculated from observed substitutions between proteins that share 62% sequence identity or less. One would use a higher numbered BLOSUM matrix for aligning two closely related sequences and a lower number for more divergent sequences.

The methods developed for sequence alignment have been applied also for structure alignment by encoding three dimensional structural information in the sequence representation. The advantage of this approach is that all powerful strategies developed for multiple sequence alignment can be directly applied to structural alignments. These methods are fully automatic, don't need initial alignments, and being very fast can compare thousands of structures available in databases. Other methods developed for structural comparisons try to maximize the match between geometrical features of the structures being compared. Comparison of structures of native and drug-resistant mutants would be helpful in the design of more effective drugs.

FUTURE DIRECTIONS

1. Comprehensively identify the structural and functional components encoded in the human genome.
2. Elucidate the organization of genetic networks and protein pathways and establish how they contribute to cellular and organismal phenotypes.
3. Develop a detailed understanding of the heritable variation in the human genome.
4. Understand evolutionary variation across species and the mechanisms underlying it.
5. Develop robust strategies for identifying the genetic contributions to disease and drug response.
6. Develop strategies to identify gene variants that contribute to good health and resistance to disease.
7. Develop genome-based approaches to prediction of disease susceptibility and drug response, early detection of illness, and molecular taxonomy of disease stages.

8. Proteomics will inevitably accelerate drug discovery, although the pace of progress in this area has been slower than was initially envisaged. An understanding of the biological networks that lie below the cell's exterior will provide a rational basis for preliminary decisions on target suitability. The proteomics of host–pathogen interactions should also be an area rich in new drug targets. Regardless of the exact format, robust mass spectrometry and protein-array platforms must be moved into clinical medicine to replace the more expensive and less reliable biochemical assays that are the basis of traditional clinical chemistry. Finally, the nascent area of chemiproteomics will not only allow mechanism of action to be discovered for many drugs, but also has the potential to resurrect innumerable failed small molecules that have dire off-target effects of unknown basis.

TOOLS AND RESOURCES FOR COMPARATIVE GENOMICS AND PROTEOMICS

Databases

This quiet revolution in the biological sciences has been enabled by our ability to collect, manage, analyze and integrate large quantities of data. In the process, bioinformatics has itself developed from something considered to be little more than information management and the creation of sequence-search tools into a vibrant field encompassing both highly sophisticated database development and active pure and applied research programmes in areas far beyond the search for sequence homology.

Databases are, with increasing sophistication, providing the scientific public with access to the data. The challenge is not collecting the data but identifying and annotating features in genomic sequence and presenting them in an intuitive fashion. The general-purpose sequence databases provide uniform access to the data and a consistent annotation for an increasing number of organisms—examples include the EMBL database, GenBank and the DNA database of Japan (DDBJ) and genome databases such as Ensembl and the National Center for Biotechnology Information (NCBI) Genome Views. Species-specific databases, such as the *Saccharomyces* Genome Database and the Mouse Genome Database, provide much richer and more complex information about individual genes. Increasingly, we are coming to realize that protein-coding genes are not the only important transcribed sequences in the genome. Rfam is a database of non-coding RNA families. Rfam provides users with covariance models—which flexibly describe the secondary structure and primary sequence consensus of an RNA sequence family as well as multiple sequence alignments representing known non-coding RNAs and provides utilities for searching sequences for their presence, including entire genomes. UniProtKB/Swiss-Prot is a curated protein sequence database which strives to provide a high level of annotation (such as the description of the function of a protein, its domains structure, post-translational modifications, variants, etc.), a minimal level of redundancy and high level of integration with other databases. The most general structural database is the Protein Data Bank (PDB). PROSITE is the database describing protein domains, families and functional sites.

Useful Servers and URL's

(1) ExPASy, (2) http://www.sbg.bio.ic.ac.uk/services.html, (3) http://genome.ucsc.edu/, (4) www.**baker**lab.org/

REFERENCES

Collins, Francis, S., Green, Eric D., Guttmacher, Alan, E. and Guyer, Mark S. (2003). *Nature*, **422:** 835-847.
Das, Amit et al. and Hosur, M.V. (2006). *Proc. Natl. Acad. Sci. USA*, **103:** 18464-18469.
Florens, L. et al. and Carussi, D. J. (2002). *Nature*, **419:** 820-826.
Feuk, L., Carson, A. R. and Scherrer, S. W. (2006). *Nature Reviews*, Genetics, **7:** 85-96.
Friedberg, I. et al. and Godzik, A. (2006). *Bioinformatics*, **23:** e219-e224.
Tyers, Mike and Mann, Matthias (2003). *Nature*, **422:** 193-197.

4 Bioinformatics—Structural Biology Interface

Desh Deepak Singh

INTRODUCTION

Bioinformatics is the field of science in which biology, computer science, and information technology merge to form a single discipline. More specifically the field conceptualizes biology in terms of physico-chemical aspects of molecules and then applies informatic techniques (maths, computer science and statistics) to understand and organize this information on a large-scale. The ultimate goal of the field is to enable the discovery of new biological insights as well as to create a global perspective from which unifying principles in biology can be discerned. At the beginning of the "genomic revolution", a bioinformatics concern was the creation and maintenance of a database to store biological information, such as nucleotide and amino acid sequences. Development of this type of database involved not only design issues but the development of complex interfaces whereby researchers could both access existing data as well as submit new or revised data. Ultimately, however, all of this information must be combined to form a comprehensive picture of normal cellular activities so that researchers may study how these activities are altered in different disease stages. Therefore, the field of bioinformatics has evolved such that the most pressing task now involves the analysis and interpretation of various types of data, including nucleotide and amino acid sequences, protein domains, and protein structures.

Some of the important milestones in the development of bioinformatics are: the first theory of molecular evolution; the Molecular Clock concept by Linus Pauling and Emile Zukerkandl in 1962; Atlas of Protein Sequences, the first protein database by Margaret Dayhoff and coworkers in 1965; Needleman-Wunsch algorithm for global protein sequence alignment in 1970; Phylogenetic taxonomy, discovery of archaea and the notion of the three primary kingdoms of life introduced by Carl Woese and co-workers in 1977; Smith-Waterman algorithm for local protein sequence alignment in 1981; the concept of a sequence motif by Russell Doolittle in 1981; Phage genome sequenced by Fred Sanger and co-workers in 1982; fast sequence similarity searching by William Pearson and David Lipman in 1985; creation of National Center for Biotechnology Information (NCBI) at NIH in 1988: BLAST: fast sequence similarity searching with rigorous statistics by Stephen Altschul, David Lipman and co-workers in 1990; first bacterial genomes completely sequenced in 1995; first archaeal and eukaryotic genome completely sequenced in 1996 and human genome (nearly) completely sequenced in 2001.

Some of the important domains of bioinformatics study are database resource generation, comparative and functional genomics, phylogeny, modeling and designing and systems biology. National Center for Biotechnology Information (NCBI), European Bioinformatics Institute (EBI), Swissprot, Sanger Research Centre, Kyoto Encyclopedia of Genes and Genomes (KEGG) and Protein Data Bank (PDB), containing information on experimentally solved structures are some of the important repositories of databases and bioinformatic tools which contain useful information on genome sequences, conserved domains, taxonomy, etc. With the rapid availability of genome sequences (available on genomes online database) of diverse organisms, analysis of the information for prediction of homologues across different species has become a very important part of bioinformatics under comparative genomics. Homologs exist amongst organisms due to a common evolutionary history and important tools have been developed which are being widely used for their identification. An important tool which is utilized for this function is BLAST (Basic local alignment search tool) which is a heuristic programme. In functional genomics efforts are made to predict function and interactions of various macromolecules in the cellular system by using available data from techniques like microarrays, etc. and extrapolating them.

An equally exciting area is the potential for uncovering evolutionary relationships and patterns between different forms of life. With the aid of nucleotide and protein sequences, it should be possible to find the ancestral ties between different organisms. Thus, so far, experience has taught us that closely related organisms have similar sequences and that more distantly related organisms have more dissimilar sequences. Proteins that show significant sequence conservation, indicating a clear evolutionary relationship, are said to be from the same protein family. By studying protein folds (distinct protein building blocks) and families, scientists are able to reconstruct the evolutionary relationship between two species and estimate the time of divergence between two organisms since they last shared a common ancestor. In phylogenetic studies, the most convenient way of visually presenting evolutionary relationships among a group of organisms is through illustrations called phylogenetic trees which can be generated using many programs like PHYLIP, MEGA, etc.

In the absence of actually determined protein structure by X-ray crystallography or nuclear magnetic resonance (NMR) spectroscopy, researchers can try to predict the three-dimensional structure using molecular modeling. This method uses experimentally determined protein structures (templates) to predict the structure of another protein that has a similar amino acid sequence (target). Identifying a protein's shape, or structure, is a key to understanding its biological function and its role in health and disease. Illuminating a protein's structure also paves the way for the development of new agents and devices to treat a disease. Yet solving the structure of a protein is no easy feat. It often takes scientists working in the laboratory months, sometimes years, to experimentally determine a single structure. Therefore, scientists have begun to turn towards computers to help predict the structure of a protein based on its sequence. The challenge lies in developing methods for accurately and reliably understanding this intricate relationship. Protein modeling involves identification of the proteins with known three-dimensional structures that are related to the target sequence, constructing a model for the target sequence based on its alignment with the template structure(s) and evaluating the model against a variety of criteria to determine if it is satisfactory. Some tools used for automated model generation are SWISS-MODEL which is available through Glaxo Wellcome Experimental Research in Geneva, Switzerland; WHAT IF available on EMBL servers and Modeller developed by Andrej Sali lab.

An important area of development in bioinformatics is in-silico designing of new therapeutics towards rational drug development and molecular interactions. The various useful tools in this arena are AutoDock (Scripps Research Institute) which is free for academia, CombiBUILD (Sandia National Labs), DockVision (University of Alberta), and DOCK (UCSF Molecular Design Institute) among many others.

The ultimate objective of all the activities in bioinformatics is to quantify the cellular interactions and wire the cell so as to say and already significant advances are being made in this direction under the general theme of systems biology (Mount, 2004; Greer, 1991; Johnson et al., 1994).

PROTEIN STRUCTURE

The protein structure function correlation is an important paradigm in understanding biology in precise molecular terms and hence makes it possible to undertake useful computational analysis and for this the understanding of the 3-D structure of proteins is very essential. A set of 20 different subunits, called amino acids, can be arranged in any order to form a polypeptide that can be thousands of amino acids long. These chains can then loop about each other or fold, in a variety of ways, but only one of these ways allows a protein to function properly. The critical feature of a protein is its ability to fold into a conformation that creates structural features, such as surface grooves, ridges and pockets, which allow it to fulfill its role in a cell. A protein's conformation is usually described in terms of levels of structure. Traditionally, proteins are looked upon as having four distinct levels of structure, with each level of structure dependent on the one below it. In some proteins, functional diversity may be further amplified by the addition of new chemical groups after synthesis is complete.

The stringing together of the amino acid chain to form a polypeptide is referred to as the primary structure. Two amino acids combine together chemically through a peptide bond, which is a chemical bond formed between the carboxyl group of one amino acid and the amino group of the second. The peptide bond is planar because it has a partial double bond and this can result in cis-trans geometric isomerism. Generally in naturally occurring proteins the trans isomer predominates. Secondary structure in proteins consists of local inter-residue interactions mediated by hydrogen bonds. The secondary structure is generated by the folding of the primary sequence and refers to the path that the polypeptide backbone of the protein follows in space. Certain types of secondary structures are relatively common. Two well-described secondary structures are alpha helix and the beta sheet. The amino acids in an α-helix are arranged in a right-handed helical structure, 5.4 Å wide. The helix has 3.6 residues per turn, and a translation of 1.5 Å along the helical axis. The amino group of an amino acid forms a hydrogen bond with the carbonyl group of the amino acid four residues earlier and this repeated hydrogen bonding defines an α-helix. β-sheets are formed when a polypeptide chain bonds with another chain that is running in the opposite direction. β-sheets may also be formed between two sections of a single polypeptide chain that is arranged such that adjacent regions are in reverse orientation. The β-sheet consists of beta strands connected laterally by three or more hydrogen bonds, forming a generally twisted, pleated sheet. The majority of β-strands are arranged adjacent to other strands and form an extensive hydrogen bond network with their neighbours in which the amino groups in the backbone of one strand establish hydrogen bonds with the carbonyl groups in the backbone of the adjacent strands.

Amino acids vary in their ability to form the various secondary structure elements. Proline and glycine are sometimes known as "helix breakers" because they disrupt the regularity of the α-helical backbone conformation; however, both have unusual conformational abilities and are commonly found in turns. Amino acids that prefer to adopt helical conformations in proteins include methionine, alanine, leucine, glutamate and lysine; by contrast, the large aromatic residues (tryptophan, tyrosine and phenylalanine) and C^{β}-branched amino acids (isoleucine, valine and threonine) prefer to adopt β-strand conformations. The tertiary structure describes the organization in three dimensions of all of the atoms in the polypeptide. If a protein consists of only one polypeptide chain, this level then describes the complete structure. Proteins have a very interesting repertoire of folds or domains into which they fold like the beta barrel, Greek key motifs, etc. In spite of the millions of protein sequences present in diverse organisms, interestingly they fold into just about 1000 different folds or domains and the study of protein folding phenomenon is a very exciting area of research engaging many researchers.

Multimeric proteins, or proteins that consist of more than one polypeptide chain, require a higher level of organization. The quaternary structure defines the conformation assumed by a multimeric protein. In this case, the individual polypeptide chains that make up a multimeric protein are often referred to as the protein subunits. The four levels of protein structure are hierarchal, that is, each level of the build process is dependent upon the one below it. A protein's primary amino acid sequence is crucial in determining its final structure. In some cases, amino acid sequence is the sole determinant, whereas in other cases, additional interactions may be required before a protein can attain its final conformation. For example, some proteins require the presence of a cofactor, or a second molecule that is part of the active protein, before it can attain its final conformation. Multimeric proteins often require one or more subunits to be present for another subunit to adopt the proper higher order structure. The entire process is cooperative, that is, the formation of one region of secondary structure determines the formation of the next region. Study of allosteric proteins is interesting because under certain conditions they have a stable alternate conformation, or shape, that enables it to carry out a different biological function. The interaction of an allosteric protein with a specific cofactor, or with another protein, may influence the transition of the protein between shapes. In addition, any change in conformation brought about by an interaction at one site may lead to an alteration in the structure, and thus function, at another site. One should bear in mind, though, that this type of transition affects only the protein's shape, not the primary amino acid sequence. Allosteric proteins play an important role in both metabolic and genetic regulation (Branden and Tooze, 1998; Chothia et al., 1977; Chothia, 1984; Lesk, 1991; Rao and Rossman, 1973 and Richardson, 1981).

PROTEIN STRUCTURE DETERMINATION

Traditionally, a protein's structure is determined using one of two techniques: X-ray crystallography or nuclear magnetic resonance (NMR) spectroscopy. Now rapid advancements in cryoelectron microscopy is also enabling determination of gross features of protein structure albeit at a low resolution.

X-ray crystallography is done on crystals of a solid form of a substance, which have an ordered array called a lattice in which the component molecules are arranged periodically. The basic building block of a crystal is called a unit cell and each unit cell contains exactly one unique set of the crystal's components, the smallest possible set that is fully representative of the crystal. When the crystal is placed in an X-ray beam, all of the unit cells present the same face to the beam; therefore, many molecules are in the same orientation with respect to the incoming X-rays. The X-ray beam enters the crystal and a number of smaller beams emerge: each one in a different direction, each one with a different intensity. If an X-ray detector, such as a piece of film, is placed on the opposite side of the crystal from the X-ray source, each diffracted ray, called a reflection, will produce a spot on the film. However, because only a few reflections can be detected with any one orientation of the crystal, an important component of any X-ray diffraction instrument is a device for accurately setting and changing the orientation of the crystal. The set of diffracted, emerging beams contains information about the underlying crystal structure.

Since its advent in the second decade of the last century, following the discovery of diffraction of X-rays by crystals in 1912 by Max von Laue, X-ray crystallography has been the method of choice for elucidating the structure of matter at the atomic and molecular levels. The wavelength of X-rays is about five thousand times less than that of light and of the same order of magnitude as the periodicity of the arrangement of atoms, ions or molecules in a crystal. Hence, crystals diffract when X-rays fall on them giving rise to intensity maxima or spots. In optical microscopy, the light waves scattered by the object are re-combined by the objective lens to produce the image. But in X-ray crystallography the re-combining has to be done by a mathematical device called Fourier synthesis. Also in X-ray diffraction, the square root of intensities gives the amplitudes of the scattered X-ray waves but information on phase angles is lost in measurement. Therefore the relative phase angles of the X-ray waves have to be determined before employing the Fourier synthesis. This is the 'phase problem' in X-ray crystallography and many methods like molecular replacement, isomorphous replacement with heavy atom derivatives and anomalous dispersion have been devised to overcome it. In the fifties, Max Perutz demonstrated that the phase problem can be solved using what is called the isomorphous replacement method in which derivative crystals are prepared by attaching heavy atoms in a coherent manner to the protein molecules in the crystal. The early protein structure solutions almost invariably employed the isomorphous replacement method which is often used in combination with what is called the anomalous dispersion method. Independent use of the anomalous dispersion method has also now gained currency. In recent decades, the molecular replacement method, which uses structural information on related proteins, is also extensively used.

One of the most important advances in the field is the development of dedicated synchrotron sources for X-ray work. In a synchrotron, an electron beam is accelerated along a long circular path which then emits electromagnetic radiation, including X-rays. The X-rays thus produced are usually several orders of magnitude more intense than those generated by conventional laboratory X-ray sources and it is also possible to choose radiation with a precise wavelength. This tunability is very useful when employing the anomalous dispersion method for structure determinations.

Biological macromolecular crystallography originated in 1934 when J.D. Bernal and Dorothy Hodgkin recorded the X-ray diffraction pattern from the crystals of pepsin. However the first structures that were actually solved were those of myoglobin and haemoglobin in the late fifties and the early sixties. John Kendrew and Max Perutz were awarded the Nobel Prize for this work in 1961. With continued technological innovations, structures of thousands of proteins have already been determined and macromolecular crystallography is now recognized as the most important component of structural biology. However a major drawback associated with X-ray crystallography technique is that crystallization of the proteins is a difficult task. Crystals are formed by slowly precipitating proteins under conditions that maintain their native conformation or structure. These exact conditions can only be discovered by repeated trials that entail varying certain experimental conditions, one at a time and this is a very time consuming and tedious process. But overall it is the technique of choice in structural biology if conditions can be optimized. It has also grown into an integral and essential part of modern biology for much of what we know now on structure-function relationships in biology at the molecular level is derived from macromolecular crystallography and has a major impact on our understanding of cellular biology and design of rational therapeutics.

Although the basic phenomenon of Nuclear Magnetic Resonance (NMR) Spectroscopy was discovered in 1938, it was in 1946 that Felix Bloch and Edward Mills Purcell refined the technique for use on liquids and solids and for which they shared the Nobel Prize in physics in 1952. Magnetic nuclei, like ^1H and ^{31}P, spin when placed in a magnetic field and these spinning nuclei can absorb radio frequency range energy specific to each nuclei. When this absorption occurs, the nucleus is described as being *in resonance* and interestingly, different atoms within a molecule *resonate* at different frequencies at a given field strength based on their chemical environment. These resonating nuclei emit a unique signal that is then picked up by a detector and processed by the Fourier Transform algorithm, a complex equation that translates the language of the nuclei into something a scientist can understand. The observation of the resonance frequencies of a molecule allows a user to discover structural information about the molecule. By measuring the frequencies at which different nuclei flip, scientists can determine molecular structure, as well as many other interesting properties of the molecule. In the past 10 years, NMR has proved to be a powerful alternative to X-ray crystallography for the determination of molecular structure. NMR has the advantage over crystallographic techniques in that experiments are performed in solution as opposed to a crystal lattice. However, the principles that make NMR possible tend to make this technique very time consuming and limit the application to small- and medium-sized molecules (Rossman and Arnold, 2006; Carter and Sweet, 1997a; 1997b, and McPherson, 2003).

India has a long tradition in crystallography. X-ray crystal structure analysis was initiated at the Indian Association for the Cultivation of Science at Kolkata in the 1930's by K. Banerjee, who was an associate of Sir C.V. Raman. Prof. G. N. Ramachandran and S. Ramaseshan, both students of Sir C.V. Raman were responsible for laying the foundations for X-ray crystallography at the Indian Institute of Science, Bangalore, in the late forties and fifties. The outstanding contributions of Prof. Ramachandran and his colleagues resulted in the 3-D structure solution of collagen and Prof. Ramachandran also made extremely useful contributions in the development of techniques of X-ray crystallography especially anomalous dispersion. The famous Ramachandran map is now text book material and epitomizes the golden era of Indian Computational Biology (Vijayan, 2007; Subramaniam, 2001 and Vijayan et al., 2000).

The bioinformatic analysis of genome sequences and gene annotation have resulted in a major data mining endeavour and some extensive structural genomic initiatives are being undertaken around the world now for 3-D structure solution of interesting targets. The structural genomic programmes of National Institute of General Medical Sciences (NIGMS) like the New York Structural Genomics Research Consortium, The Midwest Center for Structural Genomics, The TB structural genomics consortium (http://www.doe-mbi.ucla.edu/TB/), the Southeast Collaboratory for Structural Genomics and others are actively working at the interface of bioinformatics and structural biology. India is also actively participating in some of these endeavours, especially the TB structural genomics with some recent successful outcomes. Hopefully future national and international efforts in the area of bioinformatics, and structural biology will lead to a better understanding of the pathogenesis of disease and rational design of precise and specific therapeutics (Norvell and Machalek, 2000; Vijayan, 2005).

REFERENCES

Branden, C. and Tooze, J. (1998). Introduction to protein structure. Garland Publishing Inc., New York.

Carter Jr., C.W. and Sweet, R.M. (1997a). Macromolecular Crystallography, Part A. *Meth. Enzymol.*, **276**, Academic Press, San Diego.

Carter Jr., C.W. and Sweet, R.M. (1997b). Macromolecular Crystallography, Part A. *Meth. Enzymol.* **277**, Academic Press, San Diego.

Chothia, C., Levitt, M. and Richardson, D. (1977). Structure of proteins: Packing of alpha helices and beta sheets. *Proc. Natl. Acad. Sci. USA,* **74**: 4130-4134.

Chothia, C. (1984). Principles that determine the structure of proteins. *Ann. Rev. Biochem.,* **53**: 537-572.

Greer, J. (1991). Comparative modeling of homologous proteins. *Meth. Enzymol.,* **202**: 239-252.

Johnson, M.S., Srinivasan, N., Sowdhamini, R. and Blundell, T.L. (1994). Knowledge-based protein modeling. *Crit. Rev. Biochem. Mol. Biol.,* **29**: 1-68.

Lesk, A.M. (1991). Protein architecture: A practical approach. Oxford University Press, New York.

McPherson, A. (2003). Introduction to Macromolecular Crystallography. John Wiley & Sons, Hoboken, NJ.

Mount, D. (2004). Bioinformatics: Sequence and genome analysis: Cold Spring Harbor Laboratory Press, New York.

Norvell, J.C. and Machalek, A.Z. (2000). Structural genomics programs at the US National Institute of General Medical Sciences. *Nat. Struct. Mol. Biol.,* **7**: 931.

Rao, S.T. and Rossman, M.G. (1973). Comparison of super-secondary structures in proteins, *J. Mol. Biol.,* **76**: 241-256.

Richardson, J.S. (1981). The anatomy and taxonomy of protein structure. *Adv. Prot. Chem.,* **34**: 167-339.

Rossmann, M.G. and Arnold, E. (Eds.) (2006). International tables for Crystallography. Volume F: Crystallography of biological molecules. International Union of Crystallography, Chester, UK.

Subramaniam, E. (2001). G.N. Ramachandran. *Nat. Struct. Mol. Biol.,* **8**: 489-491.

Vijayan, M., Yathindra, N. and Kolaskar, A.S. (2000). Perspectives in structural biology. Universities Press (India) Ltd., Hyderabad.

Vijayan, M. (2005). Structural biology of mycobacterial proteins: The Bangalore effort. *Tuberculosis (Edinb.),* **85**: 357-366.

Vijayan, M. (2007). Peanut lectin crystallography and macromolecular structural studies in India: *J. Biosci.,* **32**: 1059.

Useful online tutorials

http://www.iucr.org/
http://www.ruppweb.org/Xray/101index.html

5 Statistical Mining of Gene and Protein Databanks

Rajani R. Joshi

CORRELATION-BASED CLUSTERING OF GENOME DATA

DNA Microarray Data

Identification and characterization of the genes and extraction of 'knowledge' from the genome databanks essentially requires annotation of promoter sequences and exons and interpretation of network relationship between gene activities. Bioinformatics research in the post-genomic era is largely focussed at this difficult task.

Of the several approaches being investigated to tackle this problem, a prominent one is the DNA-Microarray data generation and its analysis. DNA microarrays measure the expression of a gene by detecting the amount of mRNA for that gene. The mRNA is measured by tagging nucleotide sequence in a target sample and a reference sample with red and green fluorescent dyes respectively. Fluorescence intensity images are then obtained. The red and green intensities for the target and reference are measured from the images, and the ratio of red/green is then log transformed to give an expression value for each target element (Eisen et al., 1998). Positive values indicate higher expression in the target relative to the reference and negative values indicate lower expression. By comparing the expression patterns of a pair of genes in an experiment, their similarity can be characterized.

Typically, a microarray data matrix is computed from images of multicoloured bands (e.g. see Fig. 1). The horizontal strip in the image corresponds to different genes and the columns correspond to different experimental conditions. This image is quantified (in terms of intensities of different colours in the image) into a data-matrix using standard digital imaging techniques. An example of gene expression data set collected from the site (http://rana.lbl.gov/EisenData.htm) is shown in part in Table 1.

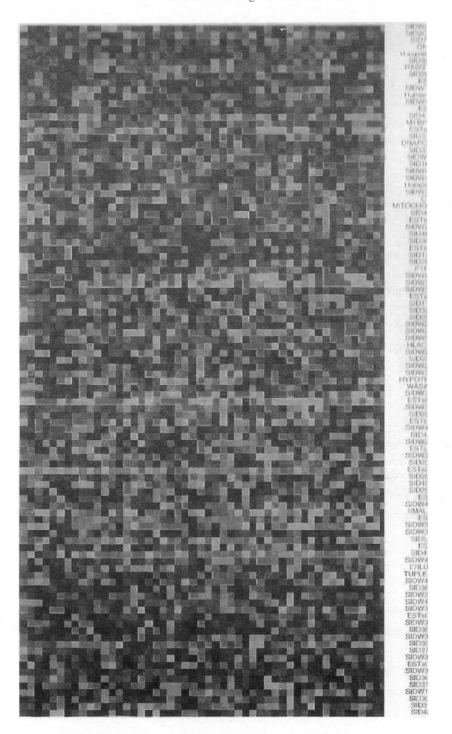

Fig. 1: Illustration of a DNA-microarray map. (Picture from
Hastie et al., 2001)

Table 1: Illustration of a DNA-microarray Data Matrix

Genes	Data points					
RPT6	0.24	−0.23	0.04	0.08	0.18	−0.22
RPN9	−0.04	−0.03	−0.03	−0.36	−0.36	0.14
BNR1	−0.17	−0.56	−0.23	0.01	0.01	−0.47
RPN10	−0.34	−0.32	−0.25	−0.4	−0.17	−0.34
PGK1	0.82	0.29	0.62	0.42	0.42	0.19

CLUSTERING TECHNIQUES

Clustering techniques are widely used for analyzing the microarray data. Clustering is a method which devices a classification scheme for grouping the objects into number of classes or clusters such that objects within the clusters are similar and dissimilar to those from any other class. Some of the methods of clustering are *supervised clustering, unsupervised clustering* and *hierarchical clustering.*

Clustering has several advantages like it is important in the areas where the relative and sequential nature of similarity between the different entities/members is also important. Also it requires less time than other methods of data mining and it can be used with larger data sets.

Correlation Based Average Linkage Hierarchical Clustering

This is a well-known technique extensively used in data mining applications including mining of the DNA *microarray* data (Eisen et al., 1998). It aims to compute a *dendrogram* that assembles all *feature vectors* in a given sample into a single tree. The leaf nodes of this tree are the individual genes in the given sample and the root node is the combined group of all of them. Starting from the leaf nodes, the most similar pairs of genes (*feature vectors*) are grouped together and placed at one node in the next layer in the successive steps.

At each step, the *data matrix* is scanned to identify the pair of most similar genes or group of genes, as the case may be. A new row is created with elements equal to the average of those along the rows being grouped. The matrix is updated with this new row replacing the grouped ones, and the process is repeated until only a single row remains. This represents the average of the *feature vectors* of the entire sample. The corresponding node (at which it would be represented in the tree) would be the 'root' node of the *dendogram* consisting of a single cluster of all the genes under study.

Implementation of Algorithm

The data matrix is constructed as described below. In each row, the missing entries, if any, are replaced by the mean of the non-missing entries. This matrix is used to calculate the similarity scores for each pair of allele in the class under consideration; this gives a square matrix whose dimension is equal to the number of allele.

The gene *similarity score matrix* is as follows:

$$\mathbf{D} = \begin{pmatrix} a_{11} & \cdots & a_{1n} \\ \vdots & \ddots & \vdots \\ a_{n1} & \cdots & a_{nn} \end{pmatrix}$$

where a_{ij} is the **correlation coefficient** $S(X_i, X_j)$ given by:

$$S(X_i, X_j) = \frac{1}{N} \sum_m \frac{\left(X_{im} -- X_{i,mean}\right)\left(X_{jm} - Y_{j,mean}\right)}{\sigma_i \qquad \sigma_j}$$

Now the entry with the maximum similarity score value is searched in the similarity matrix and the corresponding rows are returned. For example say the entry with maximum correlation is a_{ij}, then the rows i and j are returned indicating that the rows i and j are clustered. Now the rows i and j are removed and are replaced by the average of the ith and j^{th} observations. So the similarity matrix after one iteration looks like

$$\mathbf{D} = \begin{pmatrix} a^{(1)}_{11} & \cdots & a^{(1)}_{1n} \\ \vdots & \ddots & \vdots \\ a^{(1)}_{(n-1)1} & \cdots & a^{(1)}_{(n-1)(n-1)} \end{pmatrix}$$

At this stage two rows are deleted and are replaced by a single row which is the average of the observations in the deleted rows i.e. value in the ith row is added to j^{th} row column wise. This procedure is repeated till two rows are left in the similarity matrix i.e. 2×2 matrix is left. At every stage we will obtain a pair of genes that will get clubbed. Continuing in this manner we will obtain all the clustered pairs of genes at different hierarchical levels. The last two genes left in the similarity matrix will club themselves to complete the hierarchical clustering procedure.

Application

We had applied the above technique to alignment score data matrix of the VDJ genes and of the protein chains transcribed by them (data from http://imgt.cines.fr). One of the resulting *dendograms* is illustrated here in Fig. 2. The results revealed interesting properties of the variable genes which are important from understanding the role of these genes in *immunogenetic diversity* and selectivity (Joshi & Gupta, 2006).

Correlation-based hierarchical clustering by *Average-Linkage* methods are found suitable for identification of the gene families. We have extended this approach to the germline genes that generate antibodies (Igs) in our body's defense mechanism. To the best of our knowledge, no work is reported on clustering of these genes. The special structure of the data matrix based on alignment-scores is also a novel contribution, which offers potential in studying the functional association and diversity in other gene families as well.

Our results show significant roles of the non-coding portions in these genes in generation of hypervariability and also support the experimentally derived theories of multiple *exons* separated by *introns* that account for the distinct nature of the VDJ genes and diversity in the antibody specificities (Joshi & Gupta, 2006).

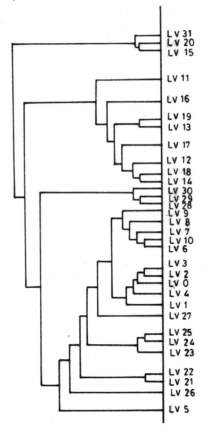

Fig. 2: Graphical drawing of the hierarchical clustering tree for Light Chain Variable Genes. The heights (from the first layer, viz., leaf nodes) of the successive layers of nodes are kept proportional to the average between group distance at the respective layers. (Figure taken from Joshi & Gupta, 2006.)

NONPARAMETRIC STATISTICS FOR PROTEIN STRUCTURE PREDICTION (The Propainor S/W)

The *PROPAINOR* (PROtein structure Prediction by AI and Nonparametric Regression) is based on nonparametric regression of the 3d-distances between residues (centroids or C_α atoms) as a function of the primary distances and some important features of the primary sequence (Joshi & Jyothi, 2003).

In this model, the unfolded polypeptide chain is represented as a linear sequence of amino acid residues, and each residue is depicted by its van der Waal's sphere. The 3d-distance between the C_α atoms of residue i and j denoted by d_{ij} is then estimated as a function of the corresponding primary distance p_{ij}, where

$$p_{ij} = r_i + 2r_{i+1} + 2r_{i+2} + \ldots + 2r_{j-1} + r_j$$

Here r_i denotes the van der Waal's radius of the ith residue.

We use a sliding window approach to extract the important parameters for given pairs (say i^{th} and j^{th}) residues in the mth window, say Wm; where, considering the *tetrahedron rule* of 3D-geometry, each window consists of successive segments of five residues say i^{th} to $(i+4)^{th}$, then $(i+1)^{th}$ to $(i+5)^{th}$ and so on; where i corresponds to i^{th} primary position on it as shown in Fig. 3.

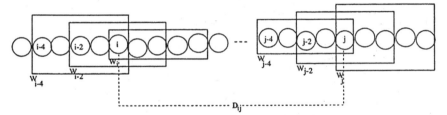

Fig. 3: Illustration of sliding window segments and the distance variable. The circles represent amino acids and the values inside the circles denote their position numbers.

Computational experiments revealed that the primary distances alone are not sufficient to explain the variation observed in the 3d-distances in native proteins. Hence, in our model, we consider a few other physical, chemical and geometrical properties of the sequence. In particular, the parameters associated with the size of the sequence, the hydrophobicity of the residues, the four stable clusters of amino acids and certain heuristics on primary-3d distance correlations were used. These parameters are found to be significant with more than 90% statistical confidence level.

Since the idea was to develop a prediction method, which did not rely on homology, the proteins in the training sample for model estimation were selected randomly from the class of proteins in the PDB having size up to 150 residues. Identical sequences were discarded. The resulting training sample had 93 proteins. The proteins in this training set were such that much less than 1% of the pairwise sequence alignments show more than 40% identity.

Estimation of Distance Constraints

The short and medium range distances $\{d_{ij} \mid j - i \leq 4\}$ are estimated as smooth additive functions of the above sequence parameters. Also, to lend compactness to the structures, certain long-range distance restraints, obtained by imposing compactness and hydrophobic core building heuristics using the theoretical results on the radius of gyration and hydrophobic residue probability distribution, are also estimated.

Performance of this method was first validated against other *ab-initio* methods that do not rely on energy minimization criteria. Our method was found to be better than other extensively used distance-based computational approaches. A comparative study of the protein structures predicted by our method with that determined by a few other distance based protein structure computation methods like DRAGON and X-PLOR showed that our method performs better in terms of the quality and accuracy of the predicted structures and also in terms of its computational efficiency (Joshi & Jyothi, 2003).

We have further extended and refined this approach so that short, medium and long range effects of the sequence are extracted in an optimal way for the estimation of short, medium and long range inter-residue C_α distance intervals. The division of the sequence is again as per the sliding window model. The short and medium range distance $\{d_{ij} \mid j - i \leq 4\}$ correlations are now discretized into three classes based on the heuristics on secondary structure. The long-range distances $\{d_{ij} \mid j - i \geq 20\}$ are discretized into two classes based on the distance being less than or greater than the average distance.

Nonparametric discriminant analysis with a Gaussian kernel on a set of sequence parameters which include local and global measures of hydrophobicity, cluster identity and secondary structure propensity gave the best cross-validated results for the estimation of the class membership in the training sample. In the case of the short and medium range distances, these estimated correlations are used in the nonparametric regression model for the estimation of the corresponding distance intervals.

The heuristic globular constraints are also refined using β-turn propensity and hydrophilicity profile plots. In particular, the residues in the sequence having β-turn propensity greater than a specified threshold or falling in the regions of local maxima of the hydrophilicity profile plots are constrained to lie on the surface of the protein.

The C_α trace of the protein is obtained by incorporating these distance interval estimates in the distance geometry algorithm, *dgsol*. The distinct structures obtained from *dgsol* are further refined using a probabilistic distance geometry program using the posterior probabilities from *nonparametric discriminant analysis*. The optimal solutions are selected based on minimal constraint violations. With this approach, the RMS of the predicted structures are found to decrease significantly—the RMS in the validation samples now vary from 4Å to 7Å (Joshi & Jyothi, 2003).

Applications: This method has successfully predicted the structure of some new proteins of which the NMR structures were found later. Noted among the pharmaceutically important proteins for which the structure is predicted using this method without and with few NMR distances is the EF-Hand Ca^{++} Binding Protein of *Entamoeba Histolytica* (Jyothi et al., 2005).

Here again the structure, in spite of the protein having two domains, is predicted with excellent accuracy ($\sim$ 6 Å when no experimental data were used; $\sim$ 4Å when about 20 NoE distance estimates were used) with the NMR predicted structure (1jfk.pdb in the protein data bank, http://www.rcsb.org/pdb).

Extension of *PROPAINO*R for longer and multi-domain proteins and its web-implementation is a part of an ongoing project sponsored by the Dept. of Biotechnology.

APPLICATION OF CART IN PROTEIN DOMAIN CLASSIFICATION

Classification and Regression Trees (*CART*) are among the most recent and extensively used statistical computational techniques suitable for Financial and Genomic Data Mining. In this project we aim to study these with special focus on applying CART to *probabilistic relational data*, in particular the protein domain data. We shall also work out some theoretical aspects of conditional probability distribution represented by CART, which is important in predictive inference on such data.

CART builds classification trees to predict categorical predictor variables and regression trees to predict continuous dependent variables.

A *Classification-tree* is generated, where we attempt to predict the class (category) of the given feature vector that has one or more continuous and/or *categorical predictor variables* as components. So for the prediction about the category, values of the predictor variables are moved through the tree until we reach a terminal node, then category for that node can be predicted. The tree is constructed from training samples using optimal splitting and stopping criteria.

Consider a simple example. Here our goal is to develop a method that can identify a person, who will buy credit card produced by a particular company. The letter B means a person is going to buy the card and N meant for a person who is not going to buy the card. This rule classifies persons as B or N depending on yes-no answers to at most three questions.

Regression Trees: A regression tree is generated, where we attempt to predict the values of a continuous variable from one or more continuous and/or categorical predictor variables. In this case also we travel down the tree to its terminal node, then the average values of the target variable present in a terminal node of the tree is the estimated value. To be more precise, if number of cases belong to the ith terminal node is n_i, and corresponding observed values of the response variables are $y_1, y_2, ..., y_{n_i}$ then $\overline{y}_i = \dfrac{1}{n_i}\sum\limits_{j=1}^{n_i} y_j$ is used

to predict value of the new cases that would belong to this terminal node.

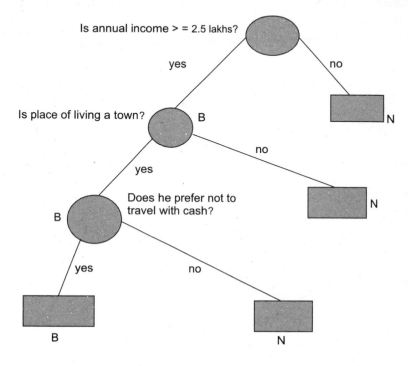

Advantages of CART

Simplicity: In most cases, the interpretation of results summarized in a tree is very simple. This simplicity is useful not only for purposes of rapid classification of new observations but can also often yield a much simpler "model" for explaining why observations are classified or predicted in a particular manner (e.g., when analyzing business problems, it is much easier to present a few simple if-then statements to management, than some elaborate equations).

Tree methods are nonparametric and nonlinear: Nowhere we are considering implicitly that the underlying relationships between the predictor variables and the dependent variable are linear, or that they are even monotonic in nature.

Selecting Splits

The split at each node is selected so as to generate the greatest improvement in predictive accuracy which is measured as some type of node impurity measure, which provides an indication of the relative homogeneity of cases in the terminal nodes. If all cases in each terminal node show identical values, then node impurity is minimal, homogeneity is maximal, and prediction is perfect (predictive validity for new cases is a different matter).

For classification problems, one may use either of several impurity measures: the Gini index, Shanon's informational entropy, Chi-square etc. By definition, it reaches a value of zero when only one class is present at a node. It reaches its maximum value when class sizes at the node are equal.

Application in Predicting the Domain Class of a Protein

We had recently applied a classification tree (*CART*) approach (Joshi, 2007) to analyze relative locations of protein domain boundaries along its length. The optimal *trees* are constructed using splits in terms of location of a *domain point* in different equal sized partitions of the protein chain. It is trained using a non-redundant sample of nearly 2000 proteins from the PDB. It has shown good accuracy of identifying domain classes for a random test subset (of over 550 proteins having less than 15% pair-wise homology): 100% correct allocation of single domain class, 78% for 2-continuous domains, 45.4% for 2-discontinuous domains, 36% for 3-continuous domains and 46.3% for 3-discontinuous domain proteins. This approach offers potential application in deciphering relational probabilistic distribution of domain boundaries conditioned on protein-length distribution and other features of the proteins sequence.

REFERENCES

Eisen, M.B., Spellman, P.T., Brown, P.O. and Botstein, D. (1998). Cluster Analysis and Display of genome-wide expression patterns. *Proc. Natl. Acad. Sci. (USA)*, **95**: 14863-14868.

Hastie, T., Tibshirani, R. and Friedman, J.H. (2001). The elements of Statistical Learning: Data Mining, Inference & Prediction. Springer Series in Statistics. Springer-Verlag, New York.

Joshi, R.R. (2007). Statistical Mining of Gene-Protein Data – CART for Structural Domains. Invited Paper at INCOB 2006, New Delhi. (Proc. in special issue of *J. Bio. Sci.*).

Joshi, R.R. and Gupta, V.K. (2006). Data Mining of VDJ Genes Reveals Interesting Clues. *Protein Peptide Letters*, **13(6)**: 587-593.

Joshi, R.R. and Jyothi, S. (2003). Ab-initio Prediction and Reliability of Protein Structural Genomics by PROPAINOR. *Computational Biology & Chemistry*, **27(3)**: 241-252.

Jyothi, S., Mustafi, S.M., Chary, K.V.R. and Joshi, R.R. (2005). Structure Prediction of a Multi-domain EF-hand Ca^{2+} Binding Protein by PROPAINOR. *J. Mol. Mod.*, **11**: 481-488.

6 Building Bioinformatic Database Systems

B.B. Meshram

INTRODUCTION

We are deluged by data—scientific data, medical data, demographic data, financial data, and marketing data. People have no time to look at this data. Human development has become a precious resource. So, we must find ways to automatically analyze data, to automatically classify it, to automatically summarize it, to automatically discover and characterize trends in it, and to automatically flag anomalies. This is one of the most active and exciting areas of the database research community. Researchers in areas such as statistics, visualization, artificial intelligence, and machine learning are contributing to this field. The breadth of the field makes it difficult to grasp its extraordinary progress over the last few years.

Bioinformatics is the application of information technology to store, organize and analyze the vast amount of biological data which is available in the form of sequences and structures of proteins (the building blocks of organisms) and nucleic acids (the information carrier). The biological information of nucleic acids is available as sequences while the data of proteins is available as sequences and structures. Sequences are represented in single dimension whereas the structure contains the three dimensional data of sequences. A biological database is a collection of data that is organized so that its contents can easily be accessed, managed, and updated. The activity of preparing a database can be divided into:

* Collection of data in a form which can be easily accessed
* Making it available to a multi-user system (always available for the user)

Databases in general can be classified into primary, secondary and composite databases. A primary database contains information of the sequence or structure alone. Examples of these include Swiss-Prot and PIR for protein sequences, GenBank and DDBJ for genome sequences and the Protein Databank for protein structures. A secondary sequence database contains information like the conserved sequence, signature sequence and active site residues of the protein families arrived by multiple sequence alignment of a set of related proteins. A secondary structure database contains entries of the PDB in an organized way. These contain entries that are classified according to their structure like all alpha proteins, all beta proteins, etc. These also contain information on conserved secondary structure motifs of a particular protein. Some of the secondary database created and hosted by various researchers at their individual laboratories include SCOP, developed at Cambridge University, CATH developed at University College of London, PROSITE of Swiss Institute of Bioinformatics, and eMOTIF at Stanford.

The sequencing of the human genome and that of other organisms is just one element of an emerging trend in the life sciences: while the knowledge, experience, and insight of researchers remain indispensable elements, the understanding of life processes is increasingly a data-driven enterprise. Huge volumes of bioinformatic data are emerging from sequencing efforts, gene expression assays, X-ray crystallography of proteins, and many other activities. The bioinformatic industry has grown so quickly that standards for the structure of data and for computational requirements have not kept pace with the data volumes. There is increasing realization, however, about the need for such standards and activity has begun to develop and promulgate them.

Problems faced due to this ever-increasing volume of data are:

- DBMS gave access to the data stored but no analysis of data.
- Analysis required to unearth the hidden relationships within the data i.e. for decision support.
- Size of databases has increased e.g. VLDBs need automated techniques for analysis as they have grown beyond manual extraction.
- The emergence of multimedia data like audio, video, images etc. and typical scientific commercial business applications.
- The bioinformatic information contain images, video, and audio data for various applications of bioinformatics.

These problems can be sorted out by implementing the bioinformatic data warehouse and internet data bases for various domains of the bioinformatics. The database server, video serrver, audio sever, image server and software tools and multimedia networking have sorted out these problems. The explosive growth in stored data such as bioinformatics has generated an urgent need for new techniques and automated tools that can intelligently assist us in transforming the vast amounts of data into useful information and knowledge. As these standards emerge Oracle has enhanced its support to the bioinformatic community by placing implementations of the standards inside the database.

This article is organized wherein the second section serves to highlight features of data base technology that support bioinformatics community. We proposed that the data modeling and process modeling can be integrated with the multimedia data base systems for the design of biological data bases. In the next section we have proposed that graphical user interface of the conventional systems can be extended to the authoring systems to collaborate the various objects of the biological databases. The subsequent section indicates that the static biological data base system can be extended to the internet biological data base for the use of distributed access of the biological data base. We can also build the data warehouse for the distribution of biological databases. The last section concludes the result and future research directions.

PROPOSED SCHEMA FOR BIOINFORMATIC DATABASES

The data model and process model will be combined together for the design of bioinformatic databases.

Data Modeling

The extended relational database provides a gradual migration path to a more object-oriented environment.

Multimedia data types: Oracle interMedia is an object relational technology that enables Oracle Database 10g to manage multimedia content (image, audio and video) in an integrated fashion with other enterprise information. The multimedia data is stored in interMedia's objects such as ORDImage, ORDVideo, ORDAudio and ORDDoc.

Relationship: Various multimedia relation types are
* is-a subclass; a is a type of b
* part-of physical part of (component) sub process of (process)

ORDImage Object Type definition using Oracle interMedia

This section contains information about digitized image concepts and using the ORDImage object type to build image applications or specialized ORDImage objects.

This object type is defined as follows in the ordispec.sql file:

```
CREATE OR REPLACE TYPE ORDImage
AS OBJECT
(
---------
- TYPE ATTRIBUTES
---------
source ORDSource,
height INTEGER,
width INTEGER,
contentLength INTEGER,
fileFormat VARCHAR2(4000),
contentFormat VARCHAR2(4000),
compressionFormat VARCHAR2(4000),
mimeType VARCHAR2(4000),
```

```
-----------
- METHOD DECLARATION
- CONSTRUCTORS

- Methods associated with image processing operations
MEMBER PROCEDURE processCopy(command IN VARCHAR2,        dest
IN OUT ORDImage),
MEMBER FUNCTION getFileFormat RETURN VARCHAR2,
MEMBER FUNCTION getContentFormat RETURN VARCHAR2,
-MEMBER FUNCTION getCompressionFormat RETURN VARCHAR2,
- Methods associated with image property set and check
  operations
- Methods associated with image attributes accessors
- Methods associated with metadata attributes
- Methods associated with the local attribute
- Methods associated with the date attribute
- Methods associated with the mimeType attribute
- Methods associated with the source attribute
..................................................................... .
- Methods associated with file operations on the source
);
```

ORDAudio Object Type definition using Oracle interMedia

This section contains information about digitized audio concepts and using the ORDAudio object type to build audio applications or specialized ORDAudio objects. This object type is defined as follows in ordaspec.sql file of interMedia:

```
CREATE OR REPLACE TYPE ORDAudio
AS OBJECT
(
— ATTRIBUTES
comments CLOB,
— AUDIO RELATED ATTRIBUTES
encoding VARCHAR2(256),
numberOfChannels INTEGER,
samplingRate INTEGER,
sampleSize INTEGER,
compressionType VARCHAR2(4000),
audioDuration INTEGER,
— METHODS
— CONSTRUCTORS
STATIC FUNCTION init( ) RETURN ORDAudio,
STATIC FUNCTION init(srcType IN VARCHAR2,
                srcLocation IN VARCHAR2,
                srcName IN VARCHAR2) RETURN ORDAudio,
— Methods associated with the date attribute
— Methods associated with the description attribute
— Methods associated with the mimeType attribute
— Methods associated with the source attribute
— Methods associated with file operations on the source
— Methods associated with audio attributes accessors
```

```
— Methods associated with setting all the properties
— Methods associated with audio processing
);
```

Implementing Video Databases using Oracle interMedia

This section contains information about digitized video concepts and using ORDVideo to build video applications or specialized ORDVideo objects. This sample object type is defined as follows in the ordvspec.sql file:

```
CREATE OR REPLACE TYPE ORDVideo
AS OBJECT
(
— ATTRIBUTES
— VIDEO RELATED ATTRIBUTES
width INTEGER,
height INTEGER,
frameResolution INTEGER,
frameRate INTEGER,
videoDuration INTEGER,
numberOfFrames INTEGER,
compressionType VARCHAR2(4000),
numberOfColors INTEGER,
bitRate INTEGER,
— METHODS
— CONSTRUCTORS
— Methods associated with the date attribute
— Methods associated with the description attribute
— Methods associated with the mimeType attribute
— Methods associated with the source attribute
— Methods associated with the video attributes accessors
MEMBER PROCEDURE setFormat(knownformat IN VARCHAR2),
MEMBER FUNCTION getFormat RETURN VARCHAR2,
MEMBER PROCEDURE getFrameSize(retWidth OUT INTEGER, retHeight
OUT INTEGER),
MEMBER PROCEDURE setFrameResolution(knownFrameResolution IN
INTEGER),
MEMBER FUNCTION getFrameResolution RETURN INTEGER,
MEMBER PROCEDURE setFrameRate(knownFrameRate IN INTEGER),
MEMBER FUNCTION getFrameRate RETURN INTEGER,
MEMBER PROCEDURE setVideoDuration(knownVideoDuration IN
INTEGER),
MEMBER FUNCTION getVideoDuration RETURN INTEGER,
MEMBER PROCEDURE setNumberOfFrames(knownNumberOfFrames IN
INTEGER),
MEMBER FUNCTION getNumberOfFrames RETURN INTEGER,
— Methods associated with setting all the properties
— Methods associated with video processing
);
```

Process Modeling

This section indicates the addition of the methods into the classes. Operations on data types for archival and retrieval are

- The input operation (insert/record) means that data will be written in the data base.
- The output (play) operations read the raw data from database.
- The modification of the image data should be done by including the following biological methods: Image Editing, Audio Editing, and Video Editing.
- Delete operations.
- Operations for queries such as search and retrieval of the data based on comparison operation etc.
- Generic biological operations in the botany and zoology should be implemented and the library should be developed, so that it will be utilised all over the world to implement the bioinformatic system. For example
- Molecular Function—elemental activity or task nuclease, DNA binding, transcription factor
- Biological Process—broad objective or goal mitosis, signal transduction, metabolism
- Cellular Component operations—location or complex nucleus, ribosome, origin recognition complex
- Gene Expression EBI Array Express, NCBI GEO

Thus data types and operations used in multimedia data bases can be integrated in both relational model and object relational data model which will give the multimedia data base schema used for bioinformatic databases.

Case studies

Case Study 1

This is a sample example case study; you can modify it as per your need.
  The person is given admission to the hospital for the operation.

```
CREATE  TABLE Person
( Admission-id       integer  PRIMARY KEY,
  Name               string
  Identification     blob
  Photoimage  ORDSYS.ORDIMAGE,
  thumb  ORDSYS.ORDIMAGE);
  Disease ()
  Medicine Prescribed ( )
  )

CREATE TABLE  Operation
( Admission-id       integer
  operation_1        ORDSYS.ORDVIDEO
  Comment            ORDSYS.ORDAUDIO
Bill( )
ReportedDocuments  ( )
  )
```

Case Study 2

This is a sample example case study; you can modify it as per your need.

```
Given  two  "parent",  how  do  we  produce  offspring?
CREATE  TABLE Father
( SSN-id              integer  PRIMARY  KEY,
  Name                string
  Identification      blob
  Photoimage ORDSYS.ORDIMAGE,
  thumb  ORDSYS.ORDIMAGE);
  Chromosome  1  string
  ....................................
  ....................................
  )
CREATE  TABLE Mother
( SSN-id              integer  PRIMARY  KEY,
  Name                string
  Identification      blob
  Photoimage ORDSYS.ORDIMAGE,
  thumb  ORDSYS.ORDIMAGE);
  Chromosome  2  string
....................................
  )
CREATE  TABLE Offspring
(id                  integer  PRIMARY  KEY,
  Chromosome  1       string
  Chromosome  2       string

Methods
Fitness  (  )
New  population  (  )
  Selection(  )
Crossover  (  )
Mutation  (  )
Accepting  (  )
Replace  (  )
Test  (  )  )  ;
```

In the above two case studies the data modeling i.e. new data types and process modeling i.e. biological methods are added into relational data base schema.

AUTHORING SYSTEM FOR BIOINFORMATIC DATABASES

Multimedia authoring systems (MAS) help users build interactive multimedia applications through automatic, or assisted, generation. Designing MAS spans a number of critical requirement issues, including the following: Hypermedia application design specifications, user interface aspects, embedding or linking the stream of object to a main document or presentation, storage of and access to multimedia objects and playing back combined streams in a synchronized manner. Multimedia applications have two components:

playback and production. Authoring systems usually provide a scripting language for specifying the playback component. The scripting language can be divided into different function classifications such as user interface, media device definition, media device control, variable definition, computations, and flow control. Device definitions explain buttons and specify media characteristics and operations. Figure 1 shows a multimedia authoring system with a fully integrated multimedia database.

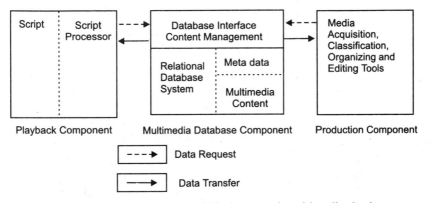

Fig. 1: Authoring system with fully integrated multimedia database.

As shown in Fig. 1, multimedia-authoring systems should include media production as well as script playback capabilities. Two options are possible: an integrated system that includes bundled utilities or a nonintegrated system in which external media tools are used. An integrated system is the best choice; such a system can also import results from external nonintegrated media products into the database. Integrated systems offer the additional advantage of seamlessly integrated production and playback. When multimedia databases become standardized, third-party products will seamlessly integrate into multimedia authoring tools.

The following steps are proposed for the authoring system of bioinformatic data.

- Planning the overall structure of application
- Planning the content of application
- Planning the interactive behaviour for user interaction with the application
- Planning the look and feel of the application
- Plan the output report of the application
- Decide the application workflow
- The input and output of the bioinformatic system should be integrated with the data modeling and process modeling and authoring system should be designed for the complete system.
- Conventional GUI should be integrated with the multimedia GUI.
- Universally recognized metaphors for the bioinformatic GUI should be built as shown in Fig. 2.

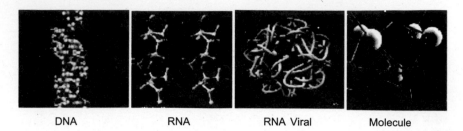

| DNA | RNA | RNA Viral | Molecule |

Fig. 2: Sample metaphors for building bioinformatic applications.

For a better understanding, the information of the application for building the bioinformatic system should be organized on the screen into five domains as shown in Fig. 3.

Menus: The pull down menu of the application should be built here.
Text attributes: All text attributes which can be inputted from the keyboard should be displayed here.
Multimedia attributes and its display in front of it: Image, audio, video attributes should be displayed here.
Buttons of the applications: To display, save, modify, edit, to switch over to other screen or to perform the application-related attributes.
Metaphors: The graphical metaphors should be defined here.

Fig. 3: Organization of the screen for designing authoring system of bioinformatics.

ADVANCED DATABASES FOR THE DISTRIBUTED BIOINFORMATICS

Members of the bioinformatics community often need to integrate data and services from different sources to build a complete solution. Because data come from different processes where each process is executed at a remote site. To enable these applications, the system needs to support a standards-based infrastructure that enables companies and their enterprise applications to communicate with other companies and their applications more efficiently. Oracle Web Services provide this infrastructure for building and deploying such applications that share data and services over the internet. Web Services consist of a set of messaging multimedia protocols, programing standards, and network registration and discovery facilities that expose business functions to authorized parties over the internet from any web connected device.

For the distribution of the information we can build the data warehouse to store the biological databases and internet biological data bases which will be useful for the decision makers and researchers in bioinformatics across the globe. These topics are beyond the scope of this article.

Data Warehouse Schema for Bioinformatics

Step 1: Build the conventional data warehouse schema
Step 2: Construct class diagram
Step 3: Map class diagram to dimension tables.
Step 4: Construct Fact Table–Primary keys and factual data about dimension tables.
Step 5: Map Object Entity ID's from Object Definition Language (ODL) to Dimensional Object ID's
Step 6: Map attributes from ODL to attributes of Dimensional ODL (DODL).
Step 7: Map method signatures from process model to method signatures of Dimensional Object Modeling (DOM).
Step 8: Map ODL model to DODL model by using step 1, step 2 and step 3.
Step 9: Design a fact table containing factual or quantitative data about the business. The object ID's of each object of DODL is a reference attribute (as foreign key in the relational databases) in the fact table.

Thus DODL contains all attributes and methods obtained from integrated object-oriented analysis for bioinformatic database schema design.

Internet Bioinformatic Databases

This program gets the XML string and prints the table data on to the standard output. In this architecture (Fig. 4) there are four tiers:

- A web browser where data and information are presented to and data are collected from the end user.
- A web server that delivers web pages, collects the data sent by the end user, and passes data to and from the application server.
- An application server that executes business rules (e.g., user authorization), formulates database queries based on the data passed by the web server, sends the queries to the back-end database, manipulates and formats the data resulting from the database query, and sends the formatted response to the web server.
- A Database server where the data are stored and managed and database requests are processed.

You can add the image server, audio server and video server depending on the complexity of your bioinformatic information.

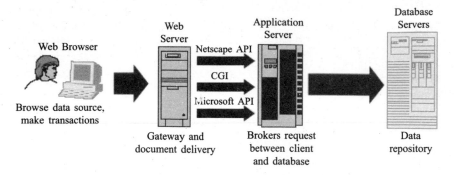

Fig. 4: Internet databases for bioinformatics.

Data Intensive XML and Oracle XDB

Oracle XDB is built upon the draft standard for XML Schema 1.0, a language used to define the structure of XML data. XDB allows the database to leverage its existing data management capabilities on any data that can be represented by an XML schema. The sample organization of the program for accessing the bioinformatics database from server to client is given in listing 1.

```
class testXMLSQL { public static void main(String[] argv)
  { try {
  // Create the JDBC Connection
  Connection conn = getConnection("scott","tiger");
 DriverManager.registerDriver(new oracle.jdbc.driver.OracleDriver());

  //XML Database Connection
  OracleXMLQuery qry = new OracleXMLQuery(conn, "select *
from person");
  String str = qry.getXMLString();
  // Print the XML output
  System.out.println(" The XML output is:\n"+str);
  qry.close();
  }catch(SQLException e) {
  System.out.println(e.toString()); } }
```

Listing 1: Program to extract the XML string from the DBMS.
The output is given below:

```
<?xml version="1.0"?>
  <ROWSET>
  <ROW id="1">
  <name>ramesh </ name >
..............................
  </ROW>
  <ROW id="2">
  ..
```

CONCLUSION AND FUTURE RESEARCH DIRECTIONS

In this work, we presented how multimedia databases can be integrated with relational system i.e. multimedia data types system to design the bioinformatic database schema. Case studies have focussed on the data base schema design of bioinformatics. The authoring system give more efficient organization and content management of rich-media data in multimedia and bioinformatic applications. The data warehouse can be useful to implement the decision support system while internet data bases can be useful to access the data globally.

The present challenge is to handle a huge volume of data, such as the ones generated by the human genome project, to improve database design, develop software for database access and manipulation, and device data-entry procedures to compensate for the varied computer procedures and systems used in different laboratories is sorted out by the proper data modeling and process modeling and authoring systems of the bioinformatic systems. But still we have to think about the security of the systems.

REFERENCES

Meshram, B.B. and Sontakke, T.R. (2001). Object Oriented Database Schema Design, 7th International Conference on Object Oriented Information Systems, Calgary, Canada, 27-29 August 2001. 497-510.

Meshram, B.B. and Sontakke, T.R. (2002). Methods of Database Schema Design. National Conference on recent trends in DBMS concepts and practice. NMAM Institute of Technology, NITTE, Udipi District, Karnataka State, India. 8-9 March 2002. 133-157.

Bansod, T.M. and Meshram, B.B. (2003.) Poster on "Biopython as a potent tool for bioinformatics", Biothiland 2003, Technology for Life, 17-20 July 2003. Pattaya, Thailand. 322

Patel, B.V. and Meshram, B.B. (2007). Carpace 1.0 for Multimedia Email Security. International Multiconference of Engineers and computer scientists. March 2007, Hong Kong (Accepted).

Bioinformatics databases

Nucleotide Databases
EMBL	www.ebi.ac.uk/embl
GenBank	www.ncbi.nlm.nih.gov/Genbank
DDBJ	www.ddbj.nig.ac.jp
DbSTS	www.ncbi.nlm.nih.gov/dbEST
Entrez	www.ncbi.nlm.nih.gov/Entrez
NCGR	www.ncgr.org/gsdb/gsdb.html
NCBI SNP	www.ncbi.nlm.nih.gov/SNP

Protein Sequence Data Bases
SWISS PROT	us.expasy.org/
TrEMBL	www.ebi.ac.uk/termbl
PIR-PSD	pir.georgetown.edu/

UniProt	www.expasy.uniprot.org
UniRef	www.ebi.ac.uk/uniref
IPI	www.ebi.ac.uk/IPI/IPIhelp.html
BLOCKS	www.blocks.fhcrc.org
PRODOM	www.sbc.su.se/~erison
EXPASY	www.expasy.ch

Protein Structure Databases

PDB	www.rcsb.org/pdb
EBI MSD	www.ebi.ac.uk/msd
NDB	ndbserver.rutgers.edu

Microarray Databases

| GEO | www.ncbi.nlm.nih.gov/geo |
| Array Express | www.ebi.ac.uk/arrayexpress |

APPENDIX A: FILE FORMATS

BMP	Bitmap image format
WMF	Windows Metafile
GIF	Graphics Interchange Format
TIFF	Tagged Image Format File
JPEG	Joint Photographers Expert Group
	-DIB file format for still images
	-Motion JEPEG Images
RIFF	Resource Image File Format
	-Wave form audio file format
	-MIDI Music Instrument Digitized interface
	-DIB Device Independent bitmapping
AVI	Audio Video Interleaved file format
HPGL	Hewlett Packard Graphics Language.
EPS	Encapsulated PostScript.
DXF	The format used by AutoCAD
CDR	The native format of CorelDraw!
PCX	ZSoft's format. Used in Corel Photopaint.

APPENDIX B: COMPRESSION

Multimedia Compression
JPEG
MPEG
H.261or Px64
Fractal Compression
DVI (Digital Video Interactive)
MHPG

APPENDIX C: PROGRAMING LANGUAGES/TOOLS

MATLAB Bioinformatics Toolbox
Java/Bio Java
Perl/ Bio-Perl
Python/Bio-Python
R/S-Plus
PHP/Bio-PHP
ORACLE 10 G –Data Base

Appendix D: Sample Implementation Hints

To successfully load the image data, you must have a mediadir directory created on your system.

```
•  —  create_mediadir.sql
•SET  SERVEROUTPUT ON;
```

- SET ECHO ON;
- CREATE OR REPLACE DIRECTORY mediadir AS 'C:/mediadir';
- — GRANT READ ON DIRECTORY mediadir TO SCOTT;

Create and Populate the *image_table* Table

- This script creates the image_table with two columns (id, image), inserts two rows, and initializes the image column.
- — create_imgtable.sql
- SET SERVEROUTPUT ON;
- SET ECHO ON;
- DROP TABLE image_table PURGE;
- CREATE TABLE image_table (id NUMBER, image ORDImage)
- — Insert rows with empty OrdImage columns and initialize the object attributes.
- INSERT INTO image_table VALUES(1,ORDImage.init());
- INSERT INTO image_table VALUES(2,ORDImage.init());
- COMMIT;

Load the Image Data

- The import_img.sql script imports image data from an image file into the ORDImage column in the image_table table using the ORDImage import( ) method. — import_img.sql
- SET SERVEROUTPUT ON;
- SET ECHO ON;
- — Import the two files into the database.
- DECLARE
- obj ORDIMAGE;
- ctx RAW(64) := NULL;
- BEGIN
- — This imports the image file img71.gif from the MEDIADIR directory
- select Image into obj from p.image_table where id = 1 for update;
- obj.setSource('file','MEDIADIR','img71.gif');
- obj.impcrt(ctx);
- update p.image_table set image = obj where id = 1;
- commit;
- — This imports the image file img50.gif from the MEDIADIR directory
- select Image into obj from image_table where id = 2 for update;
- obj.setSource('file','MEDIADIR','img50.gif');
- obj.import(ctx);
- update p.image_table set image = obj where id = 2;
- commit;
- END;

Read the Image Data from the BLOB

- The read_image.sql script : This procedure reads a specified amount of image data from the BLOB attribute, beginning at a particular offset, until all the image data is read.

```
•— Note: ORDImage has no readFromSource method like
ORDAudio , and ORDVideo; therefore, you must use the
DBMS_LOB package to read image data from a BLOB.
•— read_image.sql
•set serveroutput on
•set echo on
•create or replace procedure readimage as
•buffer RAW (32767);
•src BLOB;
•obj ORDImage;
•amt BINARY_INTEGER := 32767;
•Pos_ integer := 1;
•read_cnt integer := 1;
•BEGIN
•Select t.image.getcontent()into src fr mp.image_tablet
where t.id = 1;
•Select image into obj from image_table t where t.id = 1;
•DBMS_OUTPUT.PUT_LINE('Content length is: '|| TO_
•CHAR(obj.getContentLength())));
•LOOP
•DBMS_LOB.READ(src,amt,pos,buffer);
•DBMS_OUTPUT.PUT_LINE('start position: '|| pos);
•DBMS_OUTPUT.PUT_LINE('doing read '|| read_cnt);
•pos := pos + amt;
•read_cnt := read_cnt + 1;
•— Note: Add your own code here to process the image
data being read;
•— this routine just reads data into the buffer 32767
bytes
•— at a time, then read the next chunk, overwriting the
first
•— buffer full of data.
•END LOOP;
•EXCEPTION
•WHEN NO_DATA_FOUND THEN
•DBMS_OUTPUT.PUT_LINE('————————');
•DBMS_OUTPUT.PUT_LINE('End of data ');
•END;
•/
```

Appendix E : Help

Write the query required

```
•SAVE QUERY1
SQL>@ QUERY1
```

You will get the result of the query.
To take print out of the query result from SQL* PLUS

```
SQL>SPOOL FILENAME
SQL>WRITE QUERY AND EXECUTE
SQL>SPOOL OFF
```

Output of the query is in FILENAME.LST

To write programs in PL/SQL

1. `SQL> ED FILENAME (PRESS ENTER)`

   You will get the system editor

   Type Your Program –Procedure or Function

   `THEN SAVE IT..... ⟶ FILE..... ⟶ SAVE AS( FILENAME.SQL)`

2. To come out of the editor

   `PRESS..... ⟶ ALT F, THEN PRESS X, so that you will get SQL prompt.`

3. On this SQL prompt, for compilation give the command

   `SQL> @ FILENAME`

   This will compile your program.

4. Then to run your program, give the command

   `/ (SLASH)`

   You will get the result as procedure created.

5. To invoke a procedure in SQL* PLUS, use the command

   `SQL> EXECUTE PROCEDURENAME ( PASS PARAMETER)`

   If there is no error in the program, message will be procedure or function successfully created.

6. If errors are there, give the command

   ```
   SQL> SHOW ERROR
   SQL> L
   ```

   The program with line number will be displayed.

7. To correct your program, go to the editor and give the command

   `SQL> ED FILE NAME`

   - Make corrections in the program and save it.
   - Repeat the process to compile and run.

8. Removing server-side procedure/function

   ```
   • USING SQL*PLUS;
   SQL>DROP PROCEDURE PROCEDURE_NAME;
   SQL>DROP FUNCTION FUNCTION_NAME;
   ```

9. Creating package specification example

   ```
   SQL>CREATE OR REPLACE PACKAGE comm_package IS
   bandu-comm NUMBER := 10 ;
   PROCEDURE reset_comm.
   ( v_comm. IN NUMBER) ;
   END comm_ package ;
   ```

10. Executing a Public Procedure
   - Referencing a public variable and executing a public procedure

```
SQL>EXECUTE comm._package.g_comm.:= 5
SQL>EXECUTE comm._package.reset_comm(8)
```

11. Managing Triggers
   - Disable or re-enable a database trigger

```
ALTER TRIGGER trigger_name DISABLE ! ENABLE
```

   - Disable or re-enable ALL triggers for a table

```
ALTER TABLE table_name DISABLE ! ENABLE ALL TRIGGERS
```

   - Recompile a trigger for a table

```
ALTER TRIGGER trigger_name COMPILE
```

   - To remove trigger from database

```
DROP TRIGGER trigger_name;
```

7 Bio-sequence Signatures Using Chaos Game Representation

Achuthsankar S. Nair, Vrinda V. Nair, Arun K.S.,
Krishna Kant and Alpana Dey

INTRODUCTION

Computational biology/Bioinformatics is the application of computer sciences and allied technologies to answer the questions of biologists, about the mysteries of life. It looks as if computational biology and bioinformatics are mainly concerned with problems involving data emerging from within cells of living beings. It might be appropriate to say that computational biology and bioinformatics deal with application of computers in solving problems of molecular biology, in this context. What are these data emerging from a cell? Four important data are: DNA, RNA and Protein sequences and Micro array images. Surprisingly, first three of them are mere text data (strings, more formally) that can be opened with a text editor. The last one is a digital image which is only indirectly a cellular data (See Fig. 1).

Our interest is to discuss about deriving signatures for the first three kinds of data. It is well known that the gene regions of the DNA in the nucleus of the cell is copied (*transcribed*) into the RNA and RNA travels to protein production sites and is *translated* into proteins. In short, **DNA → RNA → Proteins**, is the central dogma of molecular biology. **Computational Genomics** and **Proteomics** are fields which encompass various studies of the genome and the proteome, based on their sequences. Both start with sequence data, and attempt to answer questions like this:

Genomics: Given a DNA sequence, where are the genes? (Gene finding); How similar is the given sequence with another one? (Pairwise sequence alignment); How similar are a set of given sequences? (Multiple sequence alignment); Where on this sequence does another given bio-molecule bind? (Transcription factor binding site identification); How can we compress this sequence? How can we visualize this sequence insightfully? (Genome browsing)

Proteomics: Given an amino acid sequence data, how similar is it with another one, or how similar are a set of amino acid sequences (Pairwise and

(a) DNA Data (4-letter strings)

(b) RNA Data (4-letter strings)

(c) Protein Data (20-letter strings)

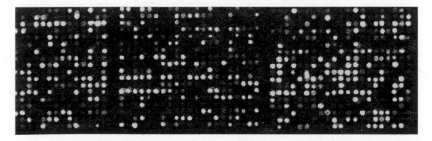

(d) Micro Array Image Data (traditional digital images)

Fig. 1: Four major kinds of data required to be analyzed in Bioinformatics.

multiple sequence alignment); What is the primary, secondary or tertiary structure of the molecule? (The great protein folding problem); Which part is most chemically active? (Active site determination problem); How would it interact with another protein? (protein-protein interaction problem); To which cell compartment is this protein belonging to? (Protein sub-cellular localization or protein sorting problem).

A large number of tools and techniques are available in computational genomics and proteomics, with varying degrees of success. A technique that has been developed successfully and used very widely in both genomics and proteomics, is the sequence alignment technique. This forms the basis for comparative studies of the genome and the proteome. A computationally intensive problem has been addressed to satisfactory level, providing a service which is quite fast and reliable.

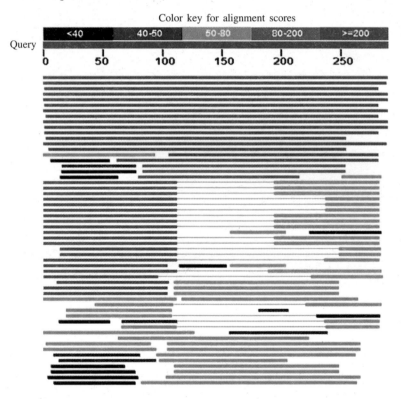

Fig. 2: Typical BLAST output for a query.

For the modern life scientist, the BLAST service which returns local alignment searches of query sequences has become the Google of biology (see Fig. 2). Another example of a successful bioinformatic tool is the UCSC genome browser. It would be unthinkable to comprehend the genomic data that is continuously erupting, without a facility such as the genome browser (see Fig. 3).

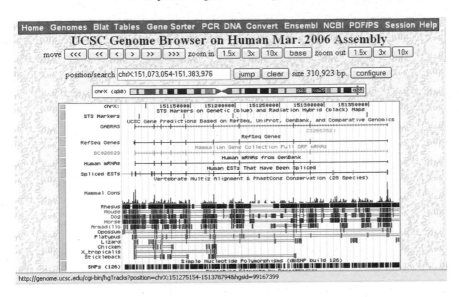

Fig. 3: The genome browser.

Many of the problems in genomics and proteomics have been attacked by various researchers using a plethora of tools from the field of mathematics, statistics, soft computing and many others. We discuss here yet another method which is useful in analyzing DNA/protein sequences aimed at solving some of the above problems.

CHAOS GAME REPRESENTATION ALGORITHM

During 1970s, a new field of physics was developed known as *chaotic dynamical systems* or simply *chaos* (Jeffrey, 1990). This field was closely associated with *fractals*. Fractal geometry, in contrast with Euclidean geometry, deals with objects that possess fractional dimensions like 1.45, 2.79 etc. Fractal geometry considers itself the geometry of the real (rather than the ideal) and consequently treats the objects in nature such as clouds, coastlines, trees, landscapes, lightning etc. as possessing fractal dimensions. Among interesting properties of the fractals are their unvarying complexity at varying scales.

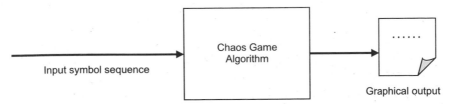

Fig. 4: CGR and its I/P and O/P.

The Chaos Game is an algorithm, which is an offshoot of research in the above area. It allows one to produce unique images of fractal nature, known as Chaos Game Representation images (CGR images) from symbolic sequences, which can serve as signature images of the sequences. It was originally described by Barnsly in 1988. Chaos Game is an algorithm whose input is a sequence of letters (finite alphabets) and output is an image (see Fig. 4).

Even though CGR of any finite sequence with finite set of alphabets is possible, here we are considering only biological sequences like DNA sequence, RNA sequence and amino acid sequences. Life scientists consider that key to life processes is centred around the above mentioned three entities. All these entities can be represented by sequences of finite alphabets.

The use of CGRs as useful signature images of bio-sequences such as DNA has been investigated since early 1990s. CGRs of genome sequences were first proposed by H. Joel Jeffrey (1990). Later other bio-sequences were also explored. We will now briefly introduce the idea of deriving a CGR image of a DNA sequence.

To derive a Chaos Game Representation of a genome, a square is first drawn to any desired scale and corners marked A, T, G and C. Points are marked within the square corresponding to the nucleotides in the sequence. In CGR, the four nucleotides A, G, C and T are assigned to the corners of a square as in Fig. 5. The choice of the corners is not based on any particular criteria, and indeed can be assigned in any other way.

Nucleotide A has an assigned position (0, 0), T has an assigned position (1, 0), G has an assigned position (1, 1) and C has an assigned position (0, 1). Now we define a procedure for representing any arbitrary nucleotide sequence as a point inside the square. For plotting a given sequence we start from the centre of the square. The first point is plotted halfway between the centre of the square, and the corner corresponding to the first nucleotide of the sequence,

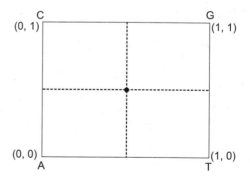

Fig. 5: CGR square with each nucleotide assigned to corner.

and successive points are plotted halfway between the previous point, and the corner corresponding to the base of each successive nucleotides. The mid point Pm (x_m, y_m) between two given points P1(x_1, y_1) and P2 (x_2, y_2) can be calculated using the following equation.

(i) $x_m = (x_1 + x_2)/2$.

(ii) $y_m = (y_1 + y_2)/2$.

These steps for plotting a given sequence are concluded below.

1. Select the first nucleotide from the given sequence.
2. Calculate the mid point between the centre and the corner corresponding to the first nucleotide $[P_N(x_N, y_N)]$. Let the mid-point be $P_i(x_i, y_i)$. Let (x_c, y_c) be the co-ordinates of the mid-point of the square.
   (a) $x_i = (x_c + x_N)/2$
   (b) $y_i = (y_c + y_N)/2$
3. Do the following steps until all the nucleotides are processed.
   (a) Read the next nucleotide in the sequence.
   (b) Calculate the mid-point between the current point $P_i(x_i, y_i)$ and the corner corresponding to the newly read nucleotide. Let the new mid-point be $P_{i+1}(x_{i+1}, y_{i+1})$.
      (i) $x_{i+1} = (x_i + x_N)/2$
      (ii) $y_{i+1} = (y_i + y_N)/2$

Now using the above procedure let us plot a DNA sequence TACAGA into this square. A square is drawn to any desired scale and corners marked A, T, G and C. Points are marked within the square corresponding to the bases in the sequence, as follows:

1. Plot the first point halfway between the centre of the square and the T corner.
2. The next point is plotted halfway between the previous point and the A corner.
3. The next point is plotted halfway between the previous point and the C corner.
4. The next point is plotted halfway between the previous point and the A corner.
5. The next point is plotted halfway between the previous point and the G corner.
6. The next point is plotted halfway between the previous point and the A corner.

Figures 6 to 11 depict the process graphically.

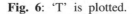

Fig. 6: 'T' is plotted.   **Fig. 7**: 'TA' is plotted.

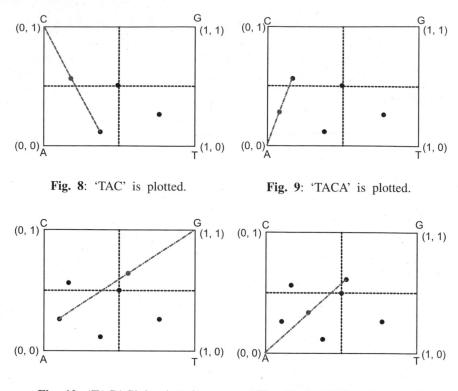

Fig. 8: 'TAC' is plotted.   Fig. 9: 'TACA' is plotted.

Fig. 10: 'TACAG' is plotted.   Fig. 11: 'TACAGA' is plotted.

Now, let us see how a real CGR would look like. Figure 12 shows CGR of Hepatitis A virus, full genome, plotted using beta version of a tool C-GRex, discussed later in this article. Figures 13 and 14 show CGR of full genome of His 2 virus, and Thermosinus carboxydivorans respectively.

Fig. 12: CGR of Hepatitis A virus full genome.

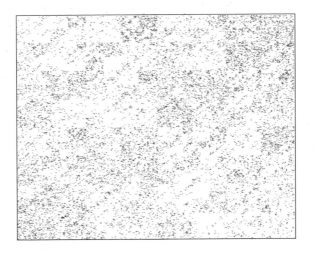

Fig. 13: CGR of His 2 virus full genome.

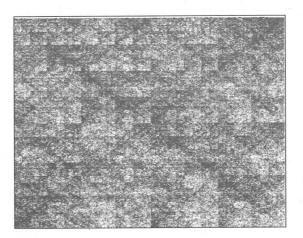

Fig. 14: CGR of Thermosinus carboxydivorans full genome.

The above CGR images clearly indicate that they vary from genome to genome, with characteristic patterns for each. This is what encourages one to investigate if CGRs can indeed be used as unique signature images for genomes and also other bio-sequences.

CGR PROPERTIES

A CGR has many properties. Every sequence has a unique CGR. In fact every symbol in a sequence will have a corresponding unique point in the CGR, even though the reverse need not be the case. Every point on the CGR is a representation of all the symbols in the sequence up to that point. For instance, in the CGR of the sequence ATTTGGCCATCG, the fifth point represents the sequence ATTTG.

Each sub-square in a CGR has a special significance. If we divide the CGR into four quadrants, then the top right corner will contain points representing sub-sequences that end with G, as a mid-point between any other point in the square and the G-corner has to fall in this quadrant. Hence if we count the points in this quadrant, it will be equal to the count of the base G in the sequence. If we divide this quadrant into another four squares, in the clockwise order, they would represent subsequences that end in GG, TG, AG and CG, making it possible to derive the 2-mer counts by counting the points in these sub-squares. In general, by dividing the CGR square into sub-squares of side 2^{-n}, we can find the number of different n-mers present in the sequence (see Fig. 15).

Side 1/2   - four sub-squares – *monomers*
Side 1/4   - 16 sub-squares – *dimers*
Side 1/8   - 64 sub-squares – *trimers* and so on

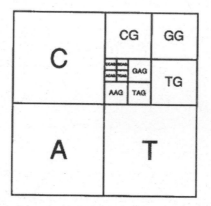

Fig. 15: Correspondence between n-mers and sub-squares of CGR.

Another interesting characteristic of CGR is that images obtained from parts of a genome show the same pattern as that of the whole genome (Deschavanne et al., 1999). Thus analysis of parts of a genome will result in a satisfactory genomic signature. This also helps comparing non-homologous genomic sequences when only parts of the genomes are available.

Genomic CGRs

The CGR of various organisms exhibit differing patterns which are characteristic of that species. Here we have obtained 20 kbp of a few species which illustrates some interesting patterns and hence features of the group. Main features of CGR images include CG double scoops, diagonals, absence of diagonals, horizontal variation in intensities from top to bottom or reverse, empty patches and word-rich regions of different shapes.

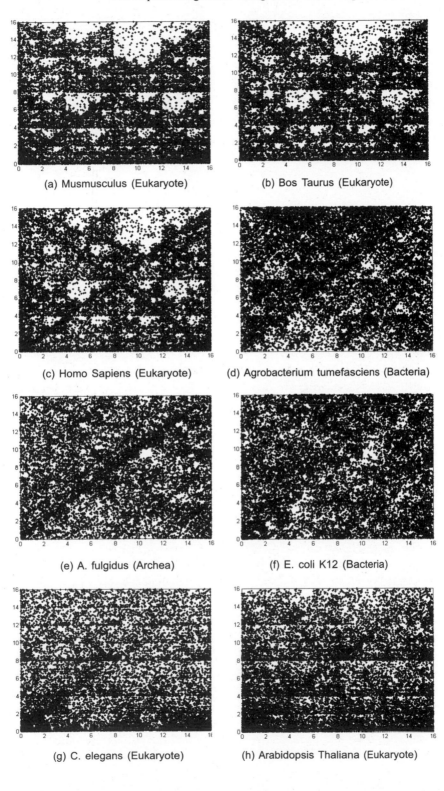

(a) Musmusculus (Eukaryote)

(b) Bos Taurus (Eukaryote)

(c) Homo Sapiens (Eukaryote)

(d) Agrobacterium tumefasciens (Bacteria)

(e) A. fulgidus (Archea)

(f) E. coli K12 (Bacteria)

(g) C. elegans (Eukaryote)

(h) Arabidopsis Thaliana (Eukaryote)

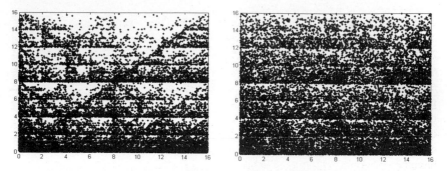

(i) Methanococcus jannaschii (Archea)   (j) Schizosaccharomyces Pombe (Eukaryote)

Fig. 16: CGRs for various organisms.

Figures 16 a, b and c are vertebrate images which exhibit CpG depletion manifesting as double scoops. Invertebrates in general have CGRs with uniform distribution of points. However certain patterns are seen as in Fig. 16 g for *C. elegans*, which shows some clustering along AG diagonal and AT line. *Agrobacterium tumefasciens* exhibit a word filled diagonal and upper triangular region whereas *Methanococcus jannaschii* and *Arabidopsis Thaliana* exhibit prominent diagonals. CpG depletion is also seen in some *eubacteria* and *archeabacteria* (see Fig. 16 i *M. jannaschii*) raising questions about the underlying mechanisms causing this depletion (Deschavanne et al., 1999). AT rich regions are exhibited in the form of horizontal lines with decreasing intensity from bottom to top by *Schizosaccharomyces Pombe* belonging to the yeast species. Referring to Fig. 15, the subsquare TAG can be still divided so that the upper left square within TAG represents CTAG. This region is seen empty in a few *eubacteria* and *archeabacteria* while the rest of the image does not show similarity. Figures 16 e and f illustrate this feature.

The double scoop in CGR image was first reported in human beta globin region. The relatively empty area corresponds to the subsquare CG and hence the inference of the relative sparseness of guanine following cytosine in the gene sequence (Jeffrey, 1990). The upper right quadrant had a large empty area, which is seen repeated in sub-quadrants presenting a *double scoop* appearance. The double scoop points out the relative sparseness of guanine following cytosine in the gene sequence. This is a simple example of using CGRs to make observations of biological relevance. Jeffrey (1990) discussed the features of CGRs of vertebrate, invertebrate, plant and slime molds, phages, bacteria and virus.

Studies on CGR of coding regions of human globin genes and alcohol dehydrogenase genes of phylogenetically divergent species were done by Hill et al. (1992). They found that CGRs were similar for genes of the same or closely related species but were different for relatively conserved genes from distantly related species. Dutta and Das (1992) reported two algorithms that can predict the presence or absence of a stretch of nucleotides in any gene family using CGRs. Nick Goldman (1993) showed that simple Markov chain models based solely on dinucleotide and trinucleotide frequencies can account for the complex pattern exhibited in CGRs of DNA sequences. Although later, Almeida et al. (2001) showed that CGR is a generalized scale independent Markov probability table. A very important and useful observation made by Deschavanne et al. (1999) was that subsequences of a genome exhibit the main characteristics of the whole genome, attesting to the validity of the genomic signature concept. Roschen et al. (2006) have explored the potential of CGR representation for making alignment-based comparisons of whole genome sequences. Classification of CGR images have been done by many researchers (Deschavanne et al., 1999; Deschavanne et al., 2000; Giron et al., 1999; Huang et al., 2004).

Proteomic CGRs

Works on CGR used for visualizing amino-acid sequences were done by Andras Fiser, Gabor E. Tusnady and Istav Simon (1994). They demonstrated that CGR can also be used for analyzing protein databases. Suggested applications include investigating regularities and motifs in the primary structure of proteins, analyzing possible structural attachments on the super-secondary structure level of proteins and testing structure prediction methods. Zu-Guo Yu, Vo Anh and Ka-Sing Lau (2004) performed multifractal and correlation analyses of the measures based on the CGR of protein sequences from complete genomes and hence attempted to construct a more precise phylogenetic tree of bacteria.

Proteomic CGR requires some tinkering with the sequence (or with the image format) to derive CGR images as amino-acids are 20 in number. We either have to group the amino-acids into four categories or change from a square to a polygon. Early approaches were based on groupings. Amino-acid properties have been well studied for long and over 500 properties are known (see AA Index, for instance). This would mean that the grouping itself would have 500 choices. If we decide to change the groupings from four to any number from 2 to 20, we can produce corresponding *n*-sided polygon images. This is a work that is being carried out by some of the authors, and is implemented in an open source tool named C-GRex, discussed next.

C-GRex: A CGR Explorer for Bio-sequences

C-GRex, Chaos Game Representation Explorer, is a tool to explore various features of CGR, in such a way that an unbelievable number of CGRs can be derived out of a given sequence, especially an amino-acid sequence, almost resembling a kaleidoscope. It is a handy tool for sequence visualization and analysis of patterns, hot spots and discoveries. C-GRex packs a wide variety of exploration facilities using Chaos Game Representation in DNA, RNA and Protein sequences and any other sequence. The tool comes with a set of functionalities which makes it unique. The software is designed in a way that a person with little knowledge about CGR will be able to work with it and manipulate the plot.

The main window of C-GRex is shown in Fig. 17. It contains three areas: plot area, settings display area, sequence panel in addition to menu bar and tool bar. The plot area shows the CGR plot of the loaded sequence. The settings applied to the CGR plot is shown in settings display area. The sequence panel contains the loaded sequence. The user can scroll through the sequence. The starting and ending number of the sequence visible in the sequence panel is displayed above the panel.

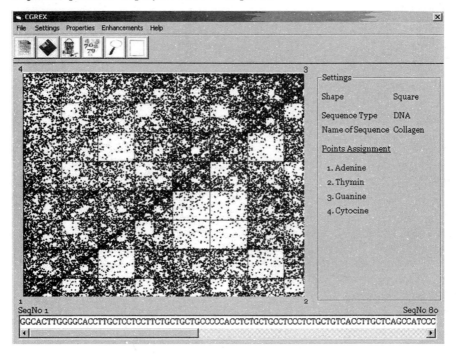

Fig. 17: The main window of C-GRex.

While it is not the intention here to introduce C-GRex comprehensively, the special feature of C-GRex will be briefly discussed. This relates to generating *n*-sided polygonal CGRs, with choice of *n* resting with the user with complete freedom to assign the corners to symbols. Let us assume that a user is interested in exploring various CGRs of given amino-acid sequence. She can choose the plot-style option in CGRex which brings up a dialog box as follows. In the space for corners, user can choose from 5 to 26 corners (3 and 4 are covered by triangle and square, 26 is set as the limit so that even plain English can be accepted by the software). See Fig. 18.

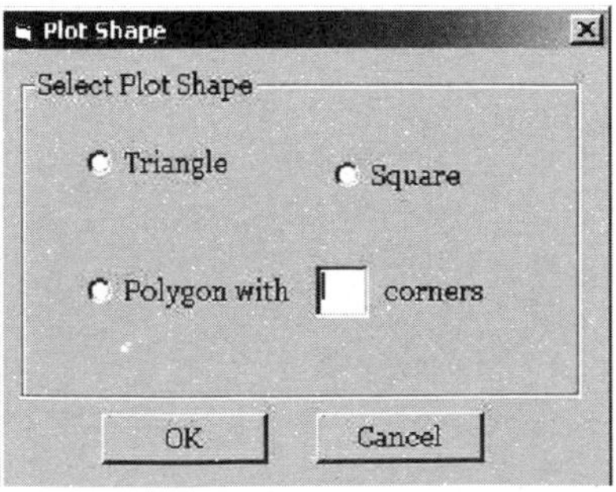

Fig. 18: The plot-style dialog box of C-GRex.

Suppose the user chooses eight corners, then 20 amino acids have to be assigned to these eight corners. C-GRex permits this to be done by a manual assignment or by an automatic assignment based on clustering and a choice of physico-chemical parameter for clustering. For clustering, it uses the well known *k-means* clustering. Figure 19 shows manual selection of amino-acid assignment for eight-cornered polygon. For clustering one of the 500 physico-chemical parameters can be chosen by the user. Figure 20 shows a eight-cornered CGR.

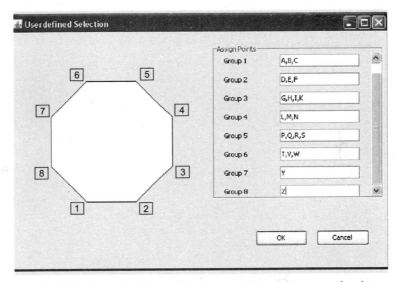

Fig. 19: Selection of amino-acid assignment for eight-cornered polygon.

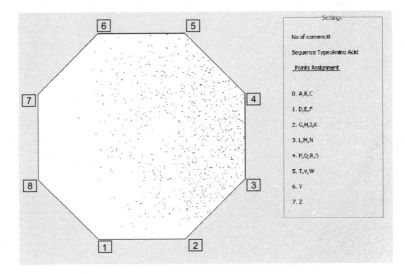

Fig. 20: Eight-cornered polygonal CGR.

CONCLUDING REMARKS

We have tried to introduce the concept and application of chaos game representation of bio-sequences. A versatile tool also was introduced. CGRs are a very different way of analyzing bio-sequences. Only a very small number of studies have been conducted in this area. The scope of CGRs is enormous. With the availability of tools such as C-GRex, it is hoped that a lot more of studies can be initiated in this area.

REFERENCES

Almeida, J.S. (2001) Analysis of genomic sequences by chaos game representation. *Bioinformatics,* **17:** 429-437.

Andras Fiser, Gabor E. Tusnady and Istvan Simon (1994). Chaos Game representation of protein structures. *J. Mol. Graphics*, **12:** 302-304.

Deschavanne, P.J., Giron, A., Vilain, J., Dufraigne, C. and Fertil, B. (2000). Genomic Signature is preserved in short DNA fragments. *IEEE.*

Deschavanne, P.J., Giron, A., Vilain, J., Fagot, G. and Fertil, B. (1999). Genomic signature: Characterization and classification of species assessed by chaos game representation of sequences. *Mol. Biol. Evol.*, **16:** 1391-1399.

Dutta, Chitra and Das, Jyotirmoy (1992) Mathematical Characterization of Chaos Game Representation New Algorithms for Nucleotide Sequence Analysis. *J Mol Biol.*, **228:** 715-719.

Giron, A., Vilain, J., Serruys, C., Brahmi, D., Deschavanne, P.J. and Fertil, B. (1999). Analysis of parametric images derived from genomic sequences using neural network based approaches. *IEEE.*

Goldman, N. (1993) Nucleotide, dinucleotide and trinucleotide frequencies explain patterns observed in chaos game representations of DNA sequences. *Nucleic Acids Res*, **21:** 2487-2491.

Hill, Kathleen, A., Schisler, Nicholas J. and Shiva M. Singh (1992) Chaos Game Representation of coding regions of human globin genes and alcohol dehydrogenase genes of phylogenetically divergent species. *J Mol Evol.*, **35:** 261-269.

Jeffrey, H.J. (1990) Chaos game representation of gene structure. *Nucleic Acids Res.* **18:** 2163–2170.

Jijoy Joseph and Roschen, Sasikumar (2006) Chaos Game Representation for comparison of whole genomes. *BMC Bioinformatics,* **7:** 243.

Xin Huang, De-Shuang Huang, Hong-Qiang Wang and Xing-Ming Zhao (2004). Representation of DNA sequences with multiple resolutions and BP neural network based classification. *IEEE.*

Zu-Guo Yu, Vo Anh and Ka-Sing Lau (2004). Chaos game representation of protein sequences based on the detailed HP model and their multifractal and correlation analyses. *J. Theor Biol.*, **226(3):** 341-348.

8 Data Mining for Bioinformatics— Microarray Data

T.V. Prasad and S.I. Ahson

INTRODUCTION

Data could be of any form, symbolic or non-symbolic, continuous or discrete, spatial or non-spatial, it should be understood that whenever the data store becomes voluminous, it requires efficient algorithms to mine out required data as well as provide methods to answer various queries. Though the data analysis techniques are useful in almost all disciplines of study, greater emphasis is given in the area of bioinformatics for mining microarray gene expression data as well as gene sequence data. Considerable work is being done in preparation of protein arrays and corresponding visualization techniques.

There are different algorithms for different situations and data forms. The algorithm that works fine for astronomical and space research cannot work for gene/protein data extraction, which in turn will be different from the one for browsing through the Internet. All techniques have different parameters/ preconditions of operation as well as advantages and disadvantages associated with them. Many require preprocessing and many do not. Many differ in dimensionality, many in terms of structure and many in conditions.

In general practice it has been observed that data matrix can be categorized into three types, viz.

- Continuous (time series), e.g., blood sugar level of a patient taken after every two hours
- Parametric (wherein each dimension or sample has some relationship with the same entity under observation), e.g., population database (covering parameters such as height, weight, eye sight, skin colour, age, hair colour, etc.) and
- Non-parametric (wherein there is no relationship between any of the samples), e.g., gene expression data, marks of students in five different subjects, etc.

The third form could be considered as a special case of second but since there are large number of applications requiring special attention towards functional dependence or independence, three different types have been enumerated here in which each sample is considered as one dimension.

DATA MINING TECHNIQUES

The complexity of data increases manifold with the increase in dimensionality (or number of features). For instance, consider the university marks processing system records having marks of 5000 students over 40 subjects (or eight semesters) as a one batch, which is already in a 2D form. The difficulty increases by one time with considering varied courses in the university as it becomes inevitable to add a new dimension or feature. Add another dimension, if there are going to be one batch per year and a group of five years is to be considered. Continue adding another dimension, if a group of universities is to be taken up. Ordinary analysis programs would take enormous time to process the data and remain a failure in answering complex queries, which could not be visualized beforehand. The existing data mining techniques could remain helpless keeping in view the voluminous amount, high dimensionality and heterogeneous nature of data. Since data have not been collected specifically for the data analysis task, often some form of preprocessing is required barring a few experimental methods. Among the various goals addressed by data preprocessing, dimensionality reduction or feature selection has been recognized as a central problem in data analysis (Talavera, 2000).

Data analysis techniques that perform well for lower dimensions fail to perform for higher dimensional data. This problem of increasing dimensions and inability of the techniques to encompass change in number of features is termed as curse of dimensionality.

With massive amounts of data lying across the world, there could be numerous techniques to take out what is required, either by artificial means or by natural processes. Most common of all these techniques are clustering and classification. The data mining tasks have been categorized as prediction methods and description methods. The prediction methods use some variables to predict unknown or future values of other variables, for instance, classification, regression and deviation detection. The description methods are used to find human-interpretable patterns that describe the data, for example, clustering, associations and classification. Techniques drawn from various other fields such as artificial intelligence, pattern recognition, statistics, database management systems and information visualization together provide efficient methods to mine the volumes (Vipin Kumar, 2002).

The data mining algorithms can be used not only for the gene expression data but for sequence data, proteomics data and pathway data as well. There are not many papers and work available for pooling data into different databases and use powerful data mining tools to mine and analyze them in the required form. Present software tools need multi-level analysis techniques to produce little meaningful results.

GENE EXPRESSION MICROARRAY TECHNOLOGY

DNA microarray is a glass slide onto which single-stranded DNA molecules are attached at fixed locations (or spots). They are composed of thousands of individual DNA sequences printed in a high-density array on a glass microscope slide using a robotic arraying device. The relative abundance of these spotted DNA sequences in two DNA or RNA samples may be assessed by monitoring the differential hybridization of the two samples (or a sample and a control) to the sequences on the array. These may be generated by single fluorescent, dual fluorescent, radioactive or colorimetric labels and the recording methods differ in each case. The arrays with the small solid substrate are also referred to as DNA chips. It is so powerful that one can investigate the gene information in short time, because at least hundreds of genes can be put on the DNA microarray to be analyzed (Cho and Won, 2003). With the currently available technologies, DNA microarrays of the capacity 61,000 genes across 22 conditions or samples can be obtained. Gene expression data indicates when and where genes are turned on (or expressed) i.e., under- or over-expression. cDNA stands for complementary DNA and is mRNA artificially translated back into DNA, but without the non-coding sequence gaps, or introns, found in the original genomic DNA.

Chee et al. (1996) made known to the world on how to use microarray data, probably when the term gene expression data was not coined. The importance of application of microarray technology to areas such as cancer diagnosis was discussed as early as 1998 (Szallasi, 1998). Bowtell (1999) discussed about various ways of obtaining gene expression data from microarray experiments. Bassett et al. (1999) highlighted about the importance and future applications of gene expression data, whereas White et al. (1999) applied microarray data to explain fruit fly metamorphosis. Golub et al. (1999) published a paper on molecular classification of cancer. This relates to their study of class discovery and class prediction by gene expression monitoring. Slonim et al. (2000) subsequently studied class prediction and discovery using gene expression data. Ramaswamy et al. (2001) brought out multi-class cancer diagnosis using tumor gene expression signature. Following types of microarrays exist (Johnny, 2002):

- DNA hybridization arrays
  - cDNA based microarrays, comes with two technologies
    - nylon membrane based - uses radioactive labeling
    - glass slides - uses fluorescence labeling
  - Oligo nucleotide microarrays for sequencing and polymorphism

- Peptide microarray chips
  - Silicon based micro-fluidics chips, 2000 to 4000 peptide sequence on a 1.5 cm^2 chip
- Protein chips
  - Secreted
  - Membranal
- Tissue arrays
- Combinatorial chemistry arrays

The problem with microarrays is perverse—a typical microarray experiment provides both too much information and too little. In most research projects, the idea is to study small number of variables and repeat the measurements over and over again. But in microarray experiments, there can be thousands of variables, corresponding to the number of individual genes being studied; but the high cost of the chips means that the number of repeated observations is usually very low. Most of the early papers only recorded whether genes were active or not, such as Li et al. (2001). The slide preparation equipment is shown in Fig. 1.

The importance of visualization of microarray gene expression data was emphasized in Carr et al. (1997), but not much research was done in the subsequent years as very few novel techniques have evolved since then. The visualization technique described here was used later in Wen et al. (1998) to demonstrate that the gene expression data can be properly mined to generate different forms of outputs such as a mutual information tree, plots of normalized time series representation, projection view in 3D stereo plot, etc. A very exhaustive collection of papers/articles on microarray technology and related allied techniques can be obtained from Leung (2002).

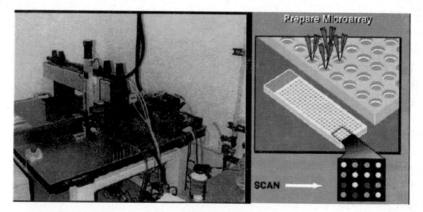

Fig. 1: The microarray slide preparation equipment consisting of robotic arm. Image source: Li et al. (2001).

Butte (2002) gave one of the detailed reviews on preparation and analysis of microarray data, while D'haeseleer et al. (1997), Wen et al. (1998) and Tibshirani et al. (1999) were the earliest reviews giving an account of great works on processing gene expression data, clustering techniques and its visualization. A wide coverage of various data mining algorithms and techniques has been given in Han and Kamber (2001).

mRNA abundance: Most popular experimental platform is used for comparing mRNA abundance in two different samples (a sample and a control). For mRNA samples, the two samples are reverse-transcribed into cDNA, labeled using different fluorescent dyes mixed (red-fluorescent dye Cy5 and green-fluorescent dye Cy3). After washing and hybridization of these samples with the arrayed DNA probes, the slides are imaged using scanner that makes fluorescence measurements for each dye. To measure the relative abundance of the hybridized RNA, the array is excited by a laser. If the RNA from the sample population is in abundance, the spot will be red; if the RNA from the control population is in abundance, it will be green. If sample and control bind equally, the spot will be yellow, while if neither binds, it will not fluoresce and appear black. Thus, from the fluorescence intensities and colours for each spot, the relative expression levels of the genes in the sample and control populations can be estimated. The log ratio between the two intensities of each dye is used as the gene expression data,

$$\text{Gene expression} = \log_2 \frac{\text{Int(Cy5)}}{\text{Int(Cy3)}} \qquad (1)$$

where Int(Cy5) and Int(Cy3) are the intensities of red and green colours. Since at least hundreds of genes are put on the DNA microarray, genome-wide information can be investigated in short time (Brazma and Vilo, 2000; Cho and Won, 2003). The main focus in genomic research is switching from sequencing to using the genome sequences in order to understand how genomes are functioning. A detailed procedure of how the gene regulation, information about the transcript levels and then understanding gene regulatory networks can be obtained from DNA microarrays (Pocock and Hubbard, 2000).

Durbin et al. (2002) described about preprocessing using statistical techniques. Yang et al. (2000 and 2001) highlighted various methods for analyzing the scanned image of the microarray gene expression data and normalization issues, so also experimental design issues for cDNA microarray were discussed by Churchill (2002) whereas Quackenbush (2002) brought out issues related to normalization of microarray data. Wong and Li (2001a and 2001b) discussed design issues, model suitability, error computation, etc. issues elaborately. An excellent description of microarray experiment preparation and data analysis through statistical methods can be found in Bergeron (2003).

Gene Expression Data Matrix or Profiles

By measuring the transcription levels of genes in an organism under various conditions, at different developmental stages and in different tissues, one can build up 'gene expression profiles' which characterize the dynamic functioning of each gene in the genome. The expression data represented in a matrix have rows representing genes, columns representing samples (e.g. various tissues, developmental stages and treatments), and each cell containing a number characterizing expression level of the particular gene in the particular sample. Such a table is called a gene expression matrix (Fig. 2, Westhead et al., 2003). Johnny (2002) in the presentation elaborated about various features of microarrays and discussed gene expression data analysis. Gene matrix could be used to compare many samples (of different creatures or patients) or of same being across different periods (i.e. time series). The samples are arranged column-wise and the gene expression value can be obtained row-wise. The first is used for comparative analysis of expression levels of genes, whereas the latter is used to determine the transformation pattern (of growth or decay) of the set of genes.

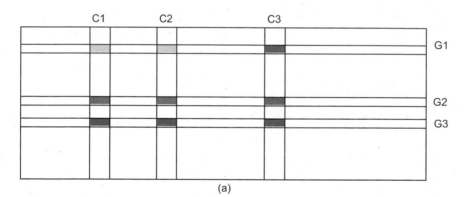

(a)

Samples → Genes↓	S1	S2	S3	S4	. . .	Sn
G1						
G2						
G3						
G4						
.						
.						
.						
Gm						

(b)

Fig. 2: Schematic of an idealized expression array, in which the results from three experiments are combined. Three genes N_G (G1, G2, G3) are labeled on vertical axis and three experimental conditions N_C (C1, C2, C3) are labeled on horizontal axis, giving a total of nine data points represented by $N_C \times N_G$. The shading of each data point represents the level of gene expression, with darker colours representing higher expression levels.

Interpretation of microarray experiment is carried out by grouping data according to similar expression profiles. It is defined as expression measurements of a given gene over a set of conditions; essentially it means reading along a row of data in the matrix. Intensity of shading is used to represent expression levels. With experimental conditions C1 and C2, genes G1 and G2 look functionally similar and G3 appears different. However, if C3 is included, a functional link between genes G1 and G3 can be seen.

Since the expression levels of most of the genes fall in more or less the same range, those genes do not seem to reveal or attach specific meaning. Such set of genes are normally avoided from further processing. One of the best examples can be found in the form of Leukemia: Acute Lymphoblastic (ALL) vs. Acute Myeloid (AML) (Golub et al., 1999), (Fig. 3).

Fig. 3: Illustration of Golub et al.'s work on ALL and AML.
Image source: Shapiro and Ramaswamy (2002).

Microarray data format: Unlike sequence and structural data, there is no international convention for the representation of data from microarray experiments. This is due to the wide variation in experimental design, assay platforms and methodologies. Recently, an initiative to develop a common language for the representation and communication of microarray data has been proposed. Experiments can be described in a standard format called MIAME and communicated using a standardized data exchange model and microarray markup language based on XML (Brazma et al., 2001; Spellman et al., 2002). Real-world microarray data is stored in a number of different formats and there are no uniform standards as of yet for representation. Broadly speaking, following are the formats in which microarray gene expression data is represented (see website of Molmine):

- GenePix
- Affymetrix
- Agilent
- ScanAnalyze
- ArrayExpress

Microarray data mining challenges: The following issues could prove to be major challenges while handling microarray gene expression data (Shapiro and Ramaswamy, 2002):

- Too few samples, usually < 100
- Too many genes, usually > 1000
- Too many columns likely to lead to false positives
- For exploration, a large set of all relevant genes is desired
- For diagnostics or identification of therapeutic targets, smallest reliable set of genes is needed
- Model needs to be explainable to biologists

Gene Expression Data Processing

Before clustering, especially with the SOM, it is essential to preprocess the data to ward off all genes with insufficient biological variations. Genes normally exhibit extraordinary level of expression and it becomes essential to normalize them across samples to limit within mean of 0 and standard deviation of 1, for each gene separately so that they can be easily plotted/ visualized. A detailed manual on gene expression pattern preprocessing by Herroro et al. (2001) lists out 10 different forms of preprocessing activities (Dopazo, 1999, 2002), though all may need not necessarily be applied to the input dataset. Different preprocessing algorithms can be applied for different types of datasets.

Apart from Hedenfalk et al. (2001), other noted works include Khan et al. (2001), Hwang et al. (2001) and Chen et al. (2000) who used machine learning techniques for diagnosis and/or prediction/profiling of breast cancer. Alizadeh et al. (2000) used gene expression data for analyzing diffuse large B-cell in lymphoma dataset. Classification of multiple tumour types was described by Yeang et al. (2001).

It is desired to have the following features searched in the microarray data:

- Understandable
- Return confidence/probability
- Robust even during presence of false positives
- Fast enough

While doing so, it must be remembered that the simple approaches are the most robust ones.

Feature reduction: Since microarray datasets are so large, classification and clustering can be labourious and demanding in terms of computer resources. It is possible to use feature reduction, where non-informative or redundant data points are removed from dataset, to make the algorithms run more quickly. For instance, if two conditions have exactly same effect on gene expression, these data are redundant and one entire column of the matrix can be eliminated. If the expression of a particular gene is same over a range of conditions, it is neither necessary nor beneficial to use this gene in further analysis because it provides no useful information on differential gene expression. In such cases, an entire row can be removed. Approaches like PCA/SVD can be used to automatically select such redundant or non-informative datasets. Redundant data are combined to form a single, composite dataset, thus reducing the dimensions of gene expression matrix and simplifying analysis (Westhead et al., 2003).

Gene Expression vs. Boolean Representation

The goal of modeling and analyzing temporal gene expression data is to determine the pattern of excitations and inhibitions between genes which make up a gene regulatory network for that organism.

One method of representation is to use Boolean networks in which gene expression values are indicated as 'on' or 'presence' and 'off' or 'absence' of specific genes. For example, the gene expression databases for multiple myeloma (an incurable cancer involving immunoglobulin secreting plasma cells) contain absolute values of gene expression involving Absent, Present and Marginal. The marginal values occuring large number of times in the datasets are ignored, thereby with only two absolute values: Absent (represented to ANNs by 0), and Present (1). Any gene that receives no 'excitation' in a time step will switch itself 'off' or 'on' in the next time step. These networks therefore model the interactions between genes in a controlled and intelligible manner. Gene expression data, including Boolean networks, are problematic because of their dimensionality (Narayanan et al., 2003).

ARTIFICIAL NEURAL NETWORKS

Numerous advantages of ANNs have been identified in both the AI and biological literature. Neural networks can perform with better accuracy than equivalent symbolic techniques (for instance decision trees) on the same data. Also, while identification tree approaches such as C4.5 can identify dominant factors the importance of which can be represented by their positions high up in the tree, ANNs may be able to detect non-dominant relationships (i.e. relationships involving several factors, each of which by itself may not be dominant) among the attributes. However, there are also a number of disadvantages. Data often has to be pre-processed to conform to the requirements of the input nodes of ANNs (e.g. normalized and converted into binary form). Training times can be long in comparison with symbolic techniques. Finally, and perhaps most importantly, solutions are encoded in the weights and therefore are not as immediately obvious as the rules and trees produced by symbolic approaches (Narayanan et al., 2003).

By preparation of large libraries of voluminous number of molecules, it became possible to identify and network the combinatorial interactions between them. A detailed study of the DNA microarray technology can be seen in Baldi and Hatfield (2001). Chakraborty (2004) gave a concise description of microarray production technology and the role of bioinformatics in processing the obtained data.

Though a number of statistical, neural network and other techniques are available, these techniques for microarray gene expression data analysis can be still considered to be in infancy stage of development (Zhang., 1999). The application of probabilistic models on array data, such as Gaussian model, Bayesian treatment and Bayesian hypothesis testing have been vividly discussed in the past (Baldi and Brunak., 2003), as well as implementation of these techniques.

Great stress on clustering and classification of DNA microarray data (or gene expression data) was laid out initially using the k-means, SOM and hierarchical clustering algorithms, which were mostly unsupervised. Classical references are DeRisi (1996), Hacia et al. (1996), Carr et al. (1997), Eisen et al. (1998), Golub et al. (1999), Tavazoie et al. (1999), Toronen et al. (2000) and Sharan et al. (2001). Description of concepts such as the use of distance measure, correlation matrix, discovering function of unknown genes, etc. could be seen in Eisen et al. (1998), being the earliest reference, whereas Carr et al. (1997) discusses about gene expression data visualization. Since then, a number of papers have become available on application of SOM, many of which clearly described on how it could be used for pattern classification than just clustering purpose.

Anderson (2001), Freeman and Skapura (1991) and Haykin (1999) presented detailed background about the ANNs and analysis of data using ANNs, whereas Baldi and Hatfield (2001) gave a summarized glance of the advances in the field of bioinformatic applications covering ANNs, gene expression data and statistics. Lesk (2002) described various tools and techniques related to proteomics, emphasizing the use of ANNs for processing protein data like protein folding. Mount (2001) brought out that ANNs have been applied to sequence data in both forms viz., genes and proteins, for promoter recognition, gene prediction and protein classification applications. Other techniques such as Multi-level perceptron (MLP), genetic algorithms (GA), radial basis function (RBF), agglomerative techniques, etc. have been sparsely used in relation to gene expression data.

In the supervised category, the most widely used algorithm is SVM, one such application can be found in Furey et al. (2000). Almost no work is available on use of LVQ with respect to gene expression data.

There are a number of lesser known ANN and non-ANN techniques available; however, very less has been employed due to various reasons such as complexity in understanding, lack of proper visualization techniques, wide acceptability, etc.

CLUSTERING AND CLASSIFICATION

Each gene in a gene expression matrix has an expression profile, that is, the expression measurements over a range of conditions (or samples). The analysis of microarray data involves grouping these data on the basis of similar expression profiles. If a predefined classification system is used to group the genes, the analysis is described as supervised. If there is no predefined classification, the analysis is described as unsupervised and is known as clustering.

Clustering first involves converting the gene expression matrix into a distance matrix, so genes with similar expression profiles can be grouped together. This generally involves calculating the Euclidean distance, the correlation measure-based distance or the Pearson linear correlation-based distance for each pair of values. Several clustering methods can then be used including hierarchical clustering, k-means clustering and the derivation of self-organizing maps. Clustering is a method (may be statistical) used to group a set of objects into a cluster of subgroups of similar objects based on a measure of similarity or "distance" between the objects. The number of clusters required must be presented before initiating the clustering process in most cases except in hierarchical clustering.

Clustering is unsupervised (i.e. anything about division is not known initially) whereas classification is a supervised learning process where division to subtypes is already known. Clustering helps in identifying the genes that are not known previously. This gives a fair idea of their function since it is presumed that a particular cluster would contain entities (or genes) having similar homology. This is similar to a classroom where students form groups or clusters automatically and based on the characteristics of one or two students in each cluster, the traits of other students can be predicted. It is believed that if not accurately predicted, at least it will give a fair idea of the traits. A comparison table of various algorithms/techniques is given in Appendix.

Separate algorithms (or modules) are needed for handling *a priori* data, for detecting outliers, for selecting distances to use, for normalizing measurements and finally, for validation/confidence estimation.

Following are some of the known classification techniques whose strong or weak features have been indicated against them (Dudoit and Gentleman, 2002a; Dudoit et al., 2000b; Thomas, 2001).

- *Decision Trees/Rules* - find smallest gene sets, but also false positives
- *Neural Nets* - works well if the number of genes is reduced
- *SVM* - good accuracy, does its own gene selection, hard to understand
- *K-nearest neighbour* - robust for small number of genes
- *Bayesian nets* - simple, robust

Additional information (or knowledge) is always needed to predict the results more precisely, for instance, while working on oncology-related datasets, the following additional items would prove to be very useful:

- There is an immanent need to combine clinical and genetic data.
- Outcome/Treatment prediction
  - ◆ Age, sex, stage of disease, etc. of the patients are useful.
  - ◆ Issues like incorrect linkages should be handled beforehand e.g. if dataset has been generated from male, it should not indicate presence of ovarian cancer.
- Molecular biologists/oncologists seem to be convinced that only a small subset of genes are responsible for particular biological properties, so they want to know which genes are most important in discriminating.
- Practical reasons, a clinical device with thousands of genes is not financially practical.
- Possible performance improvement through powerful algorithmic analysis techniques.

Goals of Clustering and Classification

It is usually thought of clustering and classification as techniques to mine the data and produce logical groups of data. However, there are other goals related to cluster analysis. Broadly speaking, various goals of cluster analysis could be laid out as the following ways:

- To find natural classes in the data
- To identify new classes/gene correlations
- To refine existing taxonomies
- To support biological analysis/discovery
- Application of different methods - hierarchical clustering, SOM's, etc.

In terms of application of ANN techniques in bioinformatics, only two techniques have been largely used viz., the SOM for the purpose of clustering data and the SVM for classification of data.

The theory of clustering has been categorized into three groups based on the measurement of distance, optimization or product (Liao, 2002):

Measurement of distance

- Distance $[d_{i,j} >= 0; \ d_{i,I} = 0; \ d_{i,j} = d_{j,I}; \ d_{i,j} <= d_{i,k} + d_{k,j}]$
- Density
- Topology
- Probability

Optimization Algorithms

- Bottom-up vs. Top-down
- Iterative vs. Recursive
- Maximum likelihood vs. Expectation maximization (Bayesian viewpoint)

Product

- Hierarchical vs. One-level
- Exclusive vs. Overlapping (clustering)
- Error tolerant (classification)
- Discrete vs. Continuous
- Generative vs. Discriminative (classification)

Baldi and Brunak (2003) brought out the prime challenges with clustering, which are as follows:

- Handling noisy signals/data could be difficult.
- Considers processing high dimensional data.
- Irregular training could lead to unstable state due to stability-plasticity dilemma.
- Clustering methods do not include *a priori* information.
- Euclidean distance and correlation similarity do not reflect co-relational structure of gene expression data; in such an event, more distance measures and correlation techniques would be needed.

Almost all the commercially as well as publicly available microarray gene expression data processing algorithms available today use SOM, *k*-means, HC and principal component analysis (PCA) techniques. Cluster analysis came to limelight in 1998, when Eisen et al. (1998) used HC and displayed genome-wide expression patterns.

The Kohonen's self-organizing maps (SOM) is the most widely used clustering algorithm, apart from *k*-means, hierarchical clustering (HC), etc. Amongst classification algorithms, Vapnik's support vector machine (SVM), decision support algorithms, reinforced learning algorithms, etc. are popular.

SELF-ORGANIZING MAPS

Tamayo et al. (1999) published about interpreting patterns of gene expression data of tumour tissues using self-organizing maps (Fig. 4). Ben-Dor et al. (1999) published a paper highlighting application of various techniques on clustering gene expression patterns. Carr et al. (1997) described about the templates for looking at gene expression clustering. In medical discipline, Wen et al. (1998) mapped the development of central nervous system using large-scale gene expression mapping. There are thousands of other applications of SOM in various fields of study, as well as in the clustering and visualization of gene expression data.

The SOM is one of the most popular clustering techniques available. It is a data visualization technique also, which reduces the dimensions of data through the use of self-organizing neural networks.

SOM are also known as topology preserving NN since the topological relationships in external stimuli are preserved and complex multi-dimensional data can be continued to be represented in the lower 2D space (Fig. 5). As observed, the serious drawbacks of SOM are that it neither produce a hierarchical classification, nor produce a proportional clustering, and the number of clusters have to be fixed beforehand i.e., in other words, SOM also requires *a priori* knowledge like any other ANN.

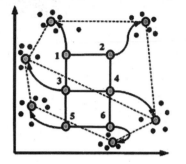
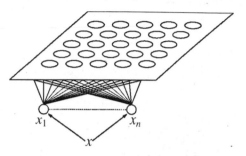

Fig. 4: The self-organizing map. Tamayo et al. (1999) discussed use of SOM on gene expression data forming them into a 3 × 2 grid of clusters.

Fig. 5: Training/learning process of neurons. Image source: Kim (2002).

Important features of SOM are:

- Kohonen's SOM is simple analog of human brain's way of organizing information in a logical manner
- Unsupervised learning process
- Arranges information in topological order
- Relationships during organization are maintained (topology preserving mapping)
- Consists of one layer of neurons in one, two or multi-dimensional arrays
- No standardized initial geometry exists
- Strongly depends on the initial inputs
- Initial learning based on training from teacher; thereafter, unsupervised
- SOM is computationally efficient algorithm located somewhere between a multi-dimensional scaling algorithm and a clustering/vector quantization algorithm
- Adaptation strength is constant over time
- New cells can be inserted and existent cells can be removed in order to adapt output map to distribution of input vectors
- Only best matching cells and its neighbourhood is adapted

The work of Iyer et al. (1999) towards application of a transcriptional program for determination of the response of human fibroblasts to serum is well noted.

SOM is a non-hierarchical clustering technique that combines visualization of output nodes also. The visualization is also known as feature map or output space. The algorithm is very similar to the adaptive *k*-means algorithm, as the selection of centroid and moving the genes (or data points) towards the nearest centroid is concerned. SOM can also be used for dimensionality reduction, as it can be considered a non-linear generalization of PCA. The steps for SOM are as follows:

- Preprocessing
  - Filter away genes with insufficient biological variation.
  - Normalize gene expression (across samples) to mean 0, standard deviation 1, for each gene separately.
- Run SOM for many iterations
- Randomly assign a data vector (a gene) to each neuron or cluster centre in the topology.
  - Find which cluster centre is nearest to the data vector.
  - All data vectors in the cluster are adjusted to move towards the cluster centre.
- Plot the results

A number of softwares have been developed to work on these areas but almost none of them cover them in great detail as they remain tilted either towards pre-processing only or mostly on visualization. SOM was also used for presenting the clustered view as visualization. The GEPAS website provides software that combines the advantages of almost all noted works on SOM together available then, and is publicly available for data submission, analysis and visualization. An interactive and visualization-rich version of SOM is presented through this work.

The number of input neurons selected depends on the specific problem, and has remained a debatable issue since the beginning. There is a different perspective on how the input should be selected and the number of input neurons be chosen. Raw data would more likely be used compared to pre-processed data. This is because one of the important features of SOM is to cluster data into classes, reducing its dimensionality. In other words, the SOM often does the required parameterization. Most works emphasize on using normalized (or pre-processed) data while working with SOM, whereas the general guideline is to normalize each input vector in its entirety. Haykin (1999) may also be referred for further details.

On the most fundamental level, normalizing weight vectors is done by dividing each weight vector component by the square root of the sum of the squares of all the weight vector components. By doing so, the total length of each weight vector from all inputs to a given output is 1. If w_{ji}' is the initial random weight generated in the interval from 0 to 1, then the normalized weight w_{ji} is given by

$$
w_{ji} = \frac{w_{ji}'}{\sqrt{\left[\sum_{i=0}^{n_i} \left(w_{ji}' \right)^2 \right]}}
\tag{2}
$$

The "neighbourhood" is the portion of the output slab (in particular, the neurons) within a specified topological radius of a given winning neurode, usually Gaussian or bubble neighbourhood. After defining the initial size of the neighbourhood, weights of all neurons in the neighbourhood of the winning neuron are adjusted. Each iteration of a complete training pattern set is a discrete step in time. Thus, the first pattern set iteration is at t_0, the next at t_1, and so on.

For the first set of iterations, the neighbourhood of the winning neuron is relatively large enough to cover the entire output slab. The neighbourhood is shrunk in size with the number of iterations, which varies with the application but is in the range of few hundred to few thousand. Picking the winning neuron is done by calculating the Euclidean distance between the input pattern vector and the weight vector associated with each output node, with the minimum Euclidean distance being the winner. Correspondingly, the radius of neighbourhood also decreases as

$$R = R_{max} \; (1 - \frac{i}{n}) \tag{3}$$

where R_{max} is the Euclidean distance between two data vectors.

To save the computing time, square root calculations are usually not done. The resulting distance calculation is defined as

$$d_{jp} = \sum_{i=1}^{n_i} \left[i_i(t) - w_{ji}(t) \right]^2 \tag{4}$$

where $d_{jp}(t)$ is the distance to neuron j for the tth iteration of pattern p. The winning neuron for the particular iteration of an input pattern is the one with the smallest distance, that is, the value of $d_{jp}(t)$. Each weight in the neighbourhood of the winning neuron is then adjusted according to equation (4). The learning coefficient $\alpha(t)$ usually decreases with increasing iterations (time).

$$w_{ji}(t+1) = w_{ji}(t) + \alpha(t) \; [i_i(t) - w_{ji}(t)] \tag{5}$$

Equations (4) and (5) are calculated for all iterations of each pattern presented to the self-organization network during training. Iterations continue until the correction in Eq. (5) becomes acceptably small (Al-Kanhal et al., 1992).

One of the drawbacks using SOM is formation of twist by the neurons during training. This can be avoided by dynamically decreasing the size of neighbourhood. Slow learning at the beginning or unstable learning at the end can also lead to differently converging solutions (Eijssen, 2000).

LEARNING VECTOR QUANTIZATION

Though old, Kohonen's learning vector quantization (LVQ) also performs well for classification of data (Kohenen et al., 1996), The LVQ algorithm has been widely applied earlier for data compression and transmission purposes. It was found to be of great importance in the field of image processing (compression, reconstruction and processing/recognition), pattern recognition, data communication through satellite, etc. and most recently anti-spam email application.

Eijssen (2000) applied various machine learning techniques and other algorithms such as fuzzy c-means and Clique partitioning algorithm for cluster analysis. The work of La Vigna (1989) spoke of non-parametric classification using LVQ, error computation, etc. aspects. Merelo and Prieto (1994) applied a variant called the G-LVQ, which is a combination of genetic algorithms and LVQ. A number of other variants such as dynamic LVQ (DLVQ), generalized relevance LVQ (GRLVQ), learning vector classification (LVC), distinction sensitive LVQ (DSLVQ) and LVQ with conscience, are also available.

The ANN based methods provide outstanding performance as compared to other methods, especially in cases where *a priori* knowledge cannot be of much help and when it becomes important for the tool to learn the characteristics of data of its own, (Kurimo, 1997). The ANNs are considered to be excellent tools in learning from local data and produce optimal output based on competitive learning.

The basic motivation behind the use of LVQ for analysis of gene expression data lies in the fact that LVQ has been used as a tool to minimize classification errors. This could become possible by stressing more on the discrimination between classes. It has been applied successfully to areas such as audio compression, data compression, data transmission, facial recognition, radar signal processing, finance and insurance, production control, sale and marketing, and so on. Keeping all these issues in view, it was felt that LVQ could be applied to such simple structured data, with higher confidence than that of SOM. One of the most amazing features of LVQ algorithm is that it can take very few vectors to obtain excellent classification results. The gene expression data provides an abstract idea of the genetic behaviour and as described above, the *a priori* knowledge is of little use. Moreover, the data presents itself void of features unlike other applications of LVQ. The gene expression data provides discrete structure instead of continuous random input as other applications do.

The unsupervised ANN technique called the Kohonen SOM could be used to transform into a supervised LVQ ANN. The basic algorithm for LVQ was introduced by Kohonen in 1986 and the technique resembles the SOM architecture, but is without a topological structure. Both SOM and LVQ replace the input vector by an approximation, which will compulsorily generate an error, and the goal here is to use LVQ for picking up the clustering algorithm with minimal expected error.

There are three variants of the LVQ algorithm, which basically guide clear discrimination of genes overlapping in different classes. Whereas the original algorithm is called the LVQ1 algorithm, LVQ2 and LVQ3 algorithms perform classification on the philosophy that if the winning neuron and runner-up neuron are at equidistance from the input vector, only then both these weight vectors should be updated.

HIERARCHICAL CLUSTERING

Dougherty et al. (2002) described application of various algorithms (including fuzzy c-means) on human fibroblast dataset published by Iyer et al. (1999). The concept of tree depth in hierarchical clustering was also described, where experiments for tree depth of 3 were considered. Vijaya et al. (2003) used Leaders-Subleaders algorithm, which is similar to the HC algorithm for clustering of protein sequence data. Of many methods for obtaining partitions of set of items, hierarchical clustering is the most used method as it produces highly elegant visual representation. Since the beginning of gene expression data mining commenced, the hierarchical clustering occupied the centre-stage along with the SOM.

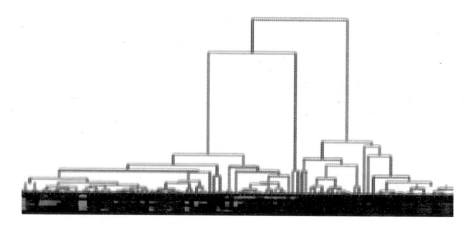

Fig. 6: The dendrogram output from hierarchical clustering.

Though it appears that the dataset is being split into different classes based on the parameters and within them subclasses and so forth, different sets of genes and experiments are combined respectively based on the distance. This combining of sets together leads to clustering of data and gives a fair idea of which set of genes is nearly related to the other sets. The output is normally a tree represented either in top-down pattern or in a left-right pattern (Fig. 6) (Eisen et al., 1998).

The hierarchical clustering has become a standard tool for cluster analysis of gene expression data as this was the first tool to be applied for microarray data analysis. The output is always connected to dendrogram (or tree) output; however, there are limitations with this output form.

There is no control within the clustering phase to ensure that the clusters are balanced. The clustered nodes when connected to form the tree provide very badly formed clusters. Probably, this is why average- or complete-linkage methods are preferred over the single-linkage method. Present tree view software do not show identical data under the same parent, as new branches are created to represent relationships even within the cluster. This cannot be true and hence, a modification to the dendrogram was suggested above. It was notable that all the methods of HC mentioned could be fused in a single formula, and parameters could be varied as needed, (Eijssen, 2000):

$$d_{k,i\cup j} = \alpha_i \, d_{ik} + \alpha_j \, d_{jk} + \beta \, d_{ij} + \delta \, |d_{ik} - d_{jk}| \qquad (6)$$

where d_{ij} indicates the (dis)similarity between cluster i and cluster j, then the clusters to be merged for each of the methods are those corresponding to the smallest updated dissimilarity or the largest updated similarity. The parameters needed for selection are given in Table 1.

Table 1: Parameters for consolidated form of HC algorithms

Method	α_i	α_j	β	δ
Nearest neighbour or single linkage	1/2	1/2	0	$-1/2$
Farthest neighbour or complete linkage	1/2	1/2	0	1/2
Average linkage	$\dfrac{|C_i|}{|C_i|+|C_j|}$	$\dfrac{|C_j|}{|C_i|+|C_j|}$	0	0
Centroid linkage	$\dfrac{|C_i|}{|C_i|+|C_j|}$	$\dfrac{|C_j|}{|C_i|+|C_j|}$	$\dfrac{-|C_i||C_j|}{\left(|C_i|+|C_j|\right)^2}$	0
Median method	1/2	1/2	$-1/4$	0
Ward's method	$\dfrac{|C_i|+|C_k|}{|C_i|+|C_j|+|C_k|}$	$\dfrac{|C_j|+|C_k|}{|C_i|+|C_j|+|C_k|}$	$\dfrac{-|C_k|}{|C_i|+|C_j|+|C_k|}$	0

Step 1: Transform (genes × experiments) matrix into (genes × genes) distance matrix. To do this, use a gene similarity metric.

$$S(X, Y) = \frac{1}{N} \sum_{i=1,N} \left(\frac{X_i - X_{\text{offset}}}{\phi_X} \right) \left(\frac{Y_i - Y_{\text{offset}}}{\phi_Y} \right)$$

$$\phi_G = \sqrt{\sum_{i=1,N} \frac{(G_i - G_{\text{offset}})^2}{N}}$$

where G_i equals the (log-transformed) primary data for gene G in condition i. For any two genes X and Y observed over a series of N conditions, G_{offset} is set to 0, corresponding to fluorescence ratio of 1.0. The equation $S(X, Y)$ is the same as Pearsons correlation except the one marked with underlined portion. This process of building a distance matrix is similar to the air distance matrix available in the table diaries.

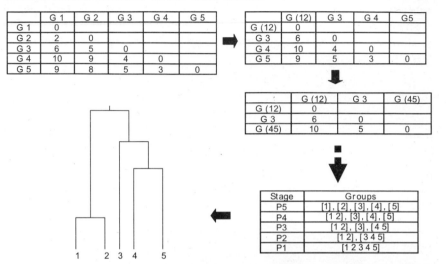

Step 2: Cluster genes based on distance matrix until single node remains.
Step 3: Once the clustering is done, use the matrix to draw a dendrogram.

The idea behind obtaining tree is to lay out the relationship between genes and cluster of genes, by means of correlation. Representation of the distance between them through extremely long branches is not preferable over the importance of relationships/linkages. The most important fact to be kept in view while using HC is to bring out the relationship between the genes, which can be nicely done using the correlation measures rather than distance metrics. Though distance metrics provides good results of clustering, the visualization will not be impressive. Another aspect of obtaining better output is the use of algorithm. In most of the cases, the complete- and average-linkage algorithms are preferred over the other two methods. The trees thus generated are more balanced and clusters nicely interpretable. At a particular stage, the Eisen's Tree View software is not preferred due to ambiguous representation of the tree (dendrogram). The output might be correct and valid, but the tree is highly cramped nearer the leaf nodes. This results in inability to visualize the output in suitable form.

k-MEANS CLUSTERING

The k-means clustering algorithm is one of the more popular clustering algorithms that has been widely adopted for biological data mining due to its simplicity of design and application. Tavazoie et al. (1999) is one such well noted implementation of k-means on gene expression data. The Stanford Biomedical Institute has a web-based application hosted for the analysis, which generates heat map or checks view output. Vijaya et al. (2003) described about the use of three variants, viz., k-means, k-medoids and k-modes. Application of an improved version of k-means algorithm that can give higher accuracy of over 93 percent was reported by Sing et al. (2003).

The algorithm supposes that the datasets are dispersed based on a distance (or similarity) matrix such as Euclidean distance in a two-dimensional space (just like the SOM).

$$d(X,\ Y) = \sqrt{\sum_{i=1}^{m}(X_i - Y_i)^2} \qquad (7)$$

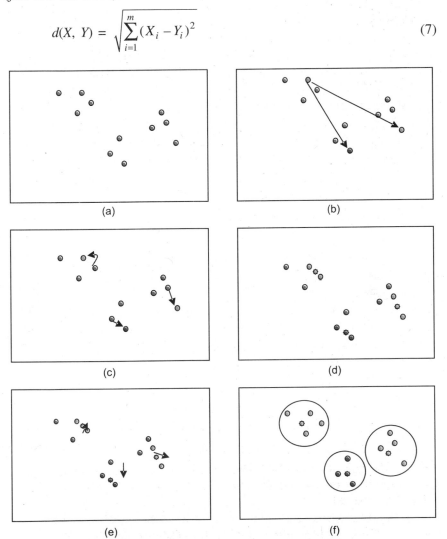

(a)

(b)

(c)

(d)

(e)

(f)

Fig. 7: k-means clustering. The k-centroids are first placed randomly and are brought to the centre of each cluster by way of average (mean) or median.

Thereafter, one cluster-centre is selected randomly and whereas the remaining k-1 cluster-centres are chosen away from all earlier centres. All the points are mapped onto the nearest cluster-centre and also the cluster-centre is slowly moved to the centre of the dataset. Graphical representation of a case is exhibited in Fig. 7. The data points were laid out on the input space, figure (a); first cluster cluster-centre or representative is randomly selected, figure (b) and farthest points are found out. The nearest data points are mapped to the representative, figure (c). Remaining k-1 representatives are identified and placed far away from the first representative and data points are mapped on to them. The algorithm is re-executed with representatives computed as an average of distance between the nearest points one after another until all are covered, figures (d) and (e). The process is repeated until no noticeable change is observed or all points have been covered, figure (f).

The k-means algorithm takes k number of random centroids and computes the means with the nearest points and all the points with minimum means would cluster around each centroid. There being three ways of adjusting the points to the centroid viz., mean, median and medoid. The mean and median methods are the two most popular methods used in k-means algorithm. Following is a brief description of the k-means algorithm:

Step 1: The distance of genes expression patterns are positioned on a 2D space based on a distance matrix. This is done by transforming n (genes) $\times m$ (experiments) matrix into n (genes) $\times n$ (genes) distance matrix.

Step 2: The first cluster centre is chosen randomly and then subsequent centres by finding the data point farthest from the centres already chosen.

Step 3: Each point is assigned to the cluster associated with the closest representative centre.

Step 4: Minimizes the within-cluster sum of squared distances from the cluster mean by moving the centroid that is computing a new cluster representative.

Step 5: Repeat steps 3 and 4 with a new representative.

Step 6: Run steps 3, 4 and 5 until no further change occurs.

A detailed description of k-means implementation can be seen in Nilsson (2002). The algorithm carefully balances the genes across various clusters; however, due to random initialization of centroid, the traceability of results cannot be ensured.

Mostly, three methods are used, viz., the k-mean, k-median and k-medoid methods. These methods differ on how the cluster centres are computed. In k-means, random cluster centres are first assigned to each cluster. All items in the proximity (minimum distance based on average) are assigned to the cluster. This process is repeated for the number of clusters chosen. The k-median algorithm is similar to the k-means except that the minimum distance is based on median of all items in the cluster, whereas k-medoid is based on the smallest sum of distance to all items in the cluster (de Hoon, 2004). The computation of minimum distance is done using the expectation minimization (EM) algorithm. k items are randomly assigned to each clusters and EM algorithm partitions the items based on optimal solution. Reassignment of an item to the nearest centroid is done by the algorithm.

PRINCIPAL COMPONENT ANALYSIS

This technique of obtaining clusters is highly useful when the data has to be partitioned on a large number of variables or features. In addition, it is most suited when it is believed that certain level of redundancy is present in those variables. The PCA algorithm has been used for dimensionality reduction. It uses a multi-stage processing for reducing the subspace by eliminating the equal expression. During the initial processing, the phase is known as SVD calculation, than in during the data normalization. Interestingly, PCA algorithm can be easily extended to visualize the analyzed knowledge in a most elegant manner. It is known as Singular Valued Decomposition (SVD) in statistical terms.

Though this method has been successfully used for few experiments, it is yet to be used extensively for mining the gene expression matrices. Alter et al. (2000) used the SVD for modeling and processing the genome-wide expression pattern of S. cerevisiae. The processing continues with N × M array of gene expression data. Corresponding to the input matrix, a matrix space M × M is created, which is represented in the L × L eigengenes space by diagonal matrix. Similar exercise is also carried out for N × N matrix space for corresponding L × L eigenarray space by diagonal matrix.

Yeung and Ruzzo (2001) used PCA for clustering gene expression data. Application of PCA for analysis of yeast sporulation time series data was taken up in Raychaudhuri et al. (2000). Wall (2003) described precise difference between PCA and SVD, and applied to gene expression analysis. The works of Wall et al. (2003) and Klingbiel (2004) have been recorded in addition to many more other publications on application of PCA/SVD on gene expression data. The description of steps involved in PCA/SVD are as follows:

Step 1: *Data input*: N-genes × M-observations

Step 2: *Perform SVD calculation:* This is done by generating matrix of dimension M × M corresponding to eigengenes as well as N × N matrix corresponding to eigenarrays.

Step 3: *Computation of pattern interference:* When these two matrix spaces are correlated, the dissimilarity or decorrelation indicate that there must be certain additional relationship between the eigengenes and eigenarrays to be established. This also indicates that these are the genes that require greater investigation.

Step 4: *Data normalization:* The decoupling of eigengenes and eigenarrays provide filtering of data without actually removing the genes from the dataset. This filtering is done by replacing all elements with zero that have corresponding expression equal to zero in the diagonal matrix thus obtained. This phase also helps in filtering out those eigengenes or eigenarrays that are responsible for noise or experimental anomalies.

Step 5: *Degenerate subspace rotation:* The matrices thus obtained are mapped/fitted onto a nearest function. This could be sine and cosine functions as depicted in the example below.

Step 6: *Data sorting:* The eigengenes can be sorted by similarity in expression level rather than by overall similarity. Correlation of any two eigengenes can be plotted along the two axes. The angular distance of each gene from the x-axis can be used to sort the dataset.

The PCA processing using GEDAS generated a detailed and much clearer visualization than other corresponding implementations as shown in Fig. 8.

The output obtained could be seen in a highly interactive manner as output corresponding to input dataset, eigen matrix, transposed matrix and the output matrix was plotted in the form of checks (dendrograph) view. It can be seen that while all other samples produced haphazard arrangement, the first two samples (rather principal components) gave interleaved pattern and exhibited certain large chunks of symmetry.

PCA is sometimes also referred to as SVD or Karhunen-Loeve (KL) expansion. PCA on one hand provides one of the best clustering and dimensionality reduction support, the only drawback it suffers from is that it can only lay linear dependencies in multivariate data. It should not be used for analysis when the outputs represent non-linear form or multiple linear dependencies. Different methods of PCA application such as supervised PCA, normalized PCA, etc. has been described by Koren and Carmel (2003).

Preprocessing Using PCA before SOM Clustering

As a preprocessing algorithm, the PCA technique was applied to the breast cancer data, which generated regular principal component vectors. All the vectors except the first two principal component vectors were deleted manually, as the first two PCs would be sufficient to reconstruct the entire input data matrix. The obtained two-dimensional matrix containing gene names, first and second PCs was preprocessed using log transform and then finally, clustered using SOM algorithm.

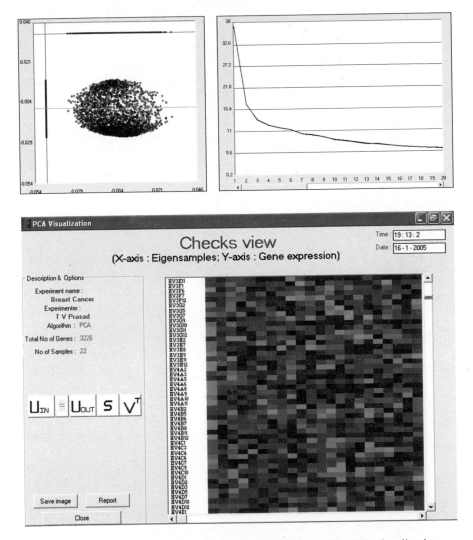

Fig. 8: The principal component visualizations: (a) The PC visualization explains how the data was clustered (or bifurcated) into two sets along the major principal component; the diagram also explained how the clusters would look once projected onto the first two PCs. (b) The eigen graph explains that the first few (say two or three) principal components would suffice for representing the entire input data matrix; and all others have lesser or no significance. (c) Full-fledged PCA analysis view; click on any of the mathematical alphabet to view the form of data matrix it points to.

Keeping the parameters standard i.e. number of clusters at 9, LR and weight at 0.5, and number of iterations at 1000, the combination of PCA separately with SOM and LVQ was worked out.

In this case, the SOM algorithm was used. For the breast cancer data, the clustering accuracy was as high as 97.74 percent. The clusters formed, as can be seen in the proximity map in Fig. 9, indicate clearly formed clusters.

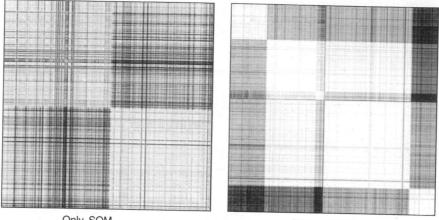

Only SOM   PCA + SOM

Fig. 9: Proximity map of SOM and LVQ outputs obtained after PCA clustering on breast cancer data.

COMPUTATION OF STANDARD ERROR

The term "standard error" is used quite often to determine the clustering/classification quality and is usually defined in terms of percentage. In statistical terms, it is known as confidence interval. The clustering error or quantization error uses Euclidean distance and is defined as

$$E = \int \|x - m_c\|^2 p(x)dx \tag{8}$$

where m_c is determined by $c = \arg\min_i \{\|x - m_i\|\}$.

In other words, the c_i is the index of centroid closest to vector i. The positions of m_c are iteratively adjusted until the factor E does not change any more. For measurement of performance of LVQ or any other clustering technique, the good old measure of mean squared error (MSE) or squared-error distortion (i.e. Euclidean distance), sometimes also known as error function or cost function, could be used:

$$MSE = \frac{1}{n}\sum_{j=0}^{n-1} d(q(x_j),\ x_j) \tag{9}$$

where $d\,(\bar{x},\ x) = \sum_{i=1}^{M}(\bar{x} - x_i)^2$.

In clustering the goal of determining error is to minimize the average (squared) error, i.e. the distance between a vector (sample) x and its representation m_{ci}. Kaski (1997) used the average squared error for clustering using k-means and SOM. Square root is usually neglected for speeding up the computational process.

Nilsson (2002) used the term "figure of merit" or FOM for defining the standard error. Considering C_1, C_2, ..., C_k being the clusters obtained, $R(g, e)$ being the expression of gene g in measurement e, and (e) being the average expression level in measurement e of genes in cluster C_i, the FOM can be defined as

$$\text{FOM }(e,\ k) = \sqrt{\frac{1}{n}\sum_{i=1}^{k}\sum_{x \in C_i}(R(x,e) - \mu_{C_i}(e))^2} \tag{10}$$

Therefore, FOM is a measure of deviation of the genes in the clustered output indicating misclassified (or missed) entities from respective cluster, over the clustering results of all genes over all measurements. The overall figure of merit is then computed as

$$\text{FOM }(k) = \sum_{e=1}^{m}\text{FOM}(e,k) \tag{11}$$

The lower the aggregate FOM, the better the clustering is considered to be. However, this measure suffers a drawback. When the number of clusters k increases, the aggregate FOM decreases. In order to compare the FOM between different clustering techniques, Yeung (2001b) proposed another measure $\sqrt{\dfrac{n-k}{n}}$ known as adjusted FOM (or AFOM), defined as

$$\text{AFOM}(e,k) = \frac{\text{FOM}(e,k)}{\sqrt{\dfrac{n-k}{n}}} \tag{12}$$

Since most of the clustering and classification algorithms use Euclidean distance as the standard metric for determining distance between the samples and their nearest centroid, the error computation also involves the Euclidean distance, as it can be seen in all noted works (Kaski, 1997). In case any other distance measure such as Pearson correlation or Manhattan distance has been used, the error function should also change correspondingly. This aspect of finding distance measure based error computation requires detailed review, especially the computation of semi-metric distance measures.

PRE-PROCESSING OF MICROARRAY GENE EXPRESSION DATA

Dopazo (1999, 2002), GEDA, Cluster, etc. software give an excellent input on the methods of pre-processing including imputation of gene expression data. Inputs from various sources was derived and incorporated in the GEDAS software.

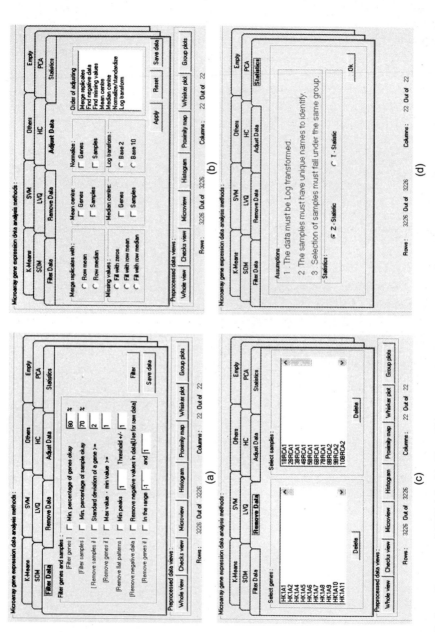

Fig. 10: Various features of GEDAS: (a) The Filter Data tab; (b) The Adjust Data tab; (c) The Remove Data tab; (d) The Statistics tab.

Loading, Filtering, Removing, Adjusting and Statistic Tests of Data

Loading Data: Before proceeding further with the experiment, the Experiment name and Experimenter name fields may be filled in promptly. The next step is to import data. Given below are the screen shots of GEDAS software related to preprocessing gene expression data. Whereas the first row specifies sample names, first column indicates the gene names. When a data file is opened for reading, the number of rows and columns should be checked and, appropriate error messages displayed in accordance with the suitability of data.

A number of options have been provided for adjusting, filtering, removing and computing various statistics of the data loaded. These functions could be accessed via the Filter Data and Adjust Data, Remove Data and Statistics tabs as shown in Fig. 10 (a) to (d).

Filtering Data: The Filter Data tab can be used to remove genes or samples that do not have certain desired properties from the dataset. While the focus should be restricted to data files and not on the scanned images, the currently available properties for data filtering are

- % Present >= X. Indicates that all genes that have missing values in greater than $(100 - X)$ percent of the columns would be removed.
- SD (Gene vector) >= X. Indicates that all samples that have standard deviations of observed *values* less than X would be removed.
- MaxVal - MinVal >= X. All genes whose differences in the gene expression values are less than X would be removed.
- Minimum peaks X with threshold ± Y. This option could be used to remove all genes that exhibit flat patterns with X number of minimum peaks falling within a threshold of ± Y. It is essential to remove non-significant genes, which are usually many in number.
- Remove negative data. This option could be used to replace the negative values with zero.

Adjusting Data: From the Adjust Data tab, a number of very important operations can be performed that alter the underlying data in the imported table. These operations include:

- **Merging duplicate data:** A number of microarray experiments have duplicate entries for genes, which may not be revealed during the manufacture of microarrays. Such entries should be removed and only one entry with suitable value should be placed. There are two methods of handling such duplicate gene data. One method is to place the mean of the gene expression values for each sample and the other is to place the median.
- **Filling the values of missing data:** During the preparation of microarray experiment or during the scanning process, or reasons otherwise, the gene

expression value for specific genes becomes unavailable, which could render the entire dataset unusable. It becomes important to fill such empty spaces with suitable values (though dummy). The best method of doing it is to fill such vacant places with zeros, row mean or row median. In some experiments, such samples are completely omitted.

- **Mean/Median Centreing:** It is very useful in experimental design where a large number of samples are compared to a reference sample. For each gene, there is a series of ratio values that are relative to the expression level of that gene in the reference sample. Since the reference sample has nothing to do with the experiment, analysis should be independent of the amount of a gene present in the reference sample. This is achieved by adjusting the values of each gene to reflect their variation from some property of the series of observed values such as the mean or median. This is what mean and/or median centreing of genes does. Centreing the data for columns/arrays can also be used to remove certain types of biases of many two-colour fluorescent hybridization experiments. Mean or median centreing the data in log-space has the effect of correcting this bias; although this is an assumption that correcting to this bias is being made. In general, the use of median rather than mean centreing is recommended, as it is more robust against outliers.

    ◆ *Mean Centre Genes and/or Arrays:* Subtract the row-wise or column-wise mean from the values in each row and/or column of data, so that mean value of each row and/or column is 0.

    ◆ *Median Centre Genes and/or Arrays:* Subtract the row-wise or column-wise mean from the values in each row and/or column of data, so that median value of each row and/or column is 0.

- **Normalize genes and/or arrays:** Normalization sets the magnitude (sum of the squares of the values) of a row/column vector to 1.0. Most of the distance metrics work with internally normalized data vectors, but the data are output as they were originally entered. If an output of normalized vectors is desired, then this option should be selected. A sample series of operations for raw data would be:
- Log transform the data
- Median centre genes and arrays
- Repeat above step five to ten times
- Normalize genes and arrays
- Repeat above step five to ten times

This results in a log-transformed, median polished (i.e. all row-wise and column-wise median values are close to zero) and normal (i.e. all row and column magnitudes are close to 1.0) dataset. Multiply all values in each row and/or column of data by a scale factor S so that the sum of the squares of the values in each row and/or column is 1.0 (a separate S is computed for each row/column).

- **Log transform data:** Replace all data values x by $\log_2(x)$. This becomes essential since all the genes have to be fitted into a frame limit. There arises a difficulty that log transformation of gene expression values usually suppresses the expression, which could be fatal to the experiment. The results of many DNA microarray experiments are fluorescent ratios. Ratio measurements are most naturally processed in log space. Whereas some researchers emphasize on log transformation, others argue that it is not required.

All the samples compulsorily follow the normal distribution of expression values as indicated in Fig. 11 for one typical sample, and in case they do provide normal distribution graph, the data must be log transformed. Data that is not log transformed and follows normal distribution would probably not require log transformation.

Removing Data: Sometimes, it is very much required to remove unknown genes from the table. This software provides the removing of genes and samples simply by selecting that particular gene or sample. These options are included in the Remove data tab.

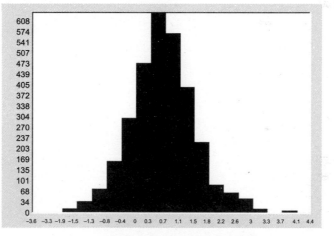

Fig. 11: The samples produce a normal distribution when frequency vs. gene expression values are plotted.

Statistics of data: In this software, two statistical tests have been included, viz., the t-statistic and z-statistic (Fig. 12 and Table 2).

When it is suspected that a particular gene x is linked to the cancer C and expression levels of this gene in n tissues with cancer C has been monitored. Further, that the monitored expression levels e_x^1, ..., e_x^n are the log-ratios of raw expression levels (of gene x in a cancerous tissue) relative to a control (the raw expression level of gene x on a non-cancerous tissue), then statistical techniques can help providing clue in the form of a set of differentially expressed genes (Jagota, 2001). Following are the techniques implemented in GEDAS:

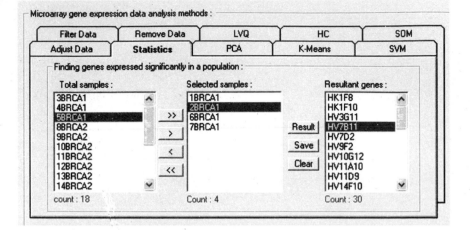

Fig. 12: *t*-statistic and *z*-statistic of gene expression data produces few significantly expressed genes in a group.

Table 2: Statistical techniques implemented in GEDAS

Sl. No.	Statistic type	Expression	Remarks
1.	*Z - Statistic*	$z(\overline{e}_x) = \dfrac{\overline{e}_x - \mu_x}{\sqrt{\sigma_x^2/n}}$	If $z(\overline{e}_x)$ is sufficiently greater than 0, then gene *x* is unregulated in the population relative to the control.
2.	*T - Statistic*	$t(\overline{e}_x) = \dfrac{\overline{e}_x - \mu_x}{s_x/\sqrt{n}}$	If $t(\overline{e}_x)$ is sufficiently greater than 0, then gene *x* is unregulated in the population relative to the control.

Distance Measures

Distances are measured using distance functions, which follow triangle inequality i.e.

$$d\,(u,\,v) \le d\,(u,\,w) + d\,(w,\,v) \tag{1.13}$$

In general language, inequality means the shortest distance between two points as a straight line. None of the correlation coefficient-based distance functions satisfy the triangle inequality and hence are known as semi-metric.

Computation of distance measure (sometimes also referred to as similarity or dissimilarity measure) is one such new feature. From the existing eight measures, the number of distance measures has been enhanced to 19 in GEDAS. The distance measures (Jagota, 2001), implemented in the GEDAS software are listed in Table 3.

Table 3: Distance measures implemented in GEDAS

Sl. No.	Distance or similarity measure	Expression Remarks
1.	Euclidean distance	$d = \sum_{i=1}^{n} (x_i - y_i)^2$

It is a true metric, as it satisfies the triangle inequality, and is the most widely used distance measure of all available. In this formula, the expression data x_i and y_i are subtracted directly from each other. We should therefore make sure that the expression data are properly normalized when using the Euclidean distance, for example by converting the measured gene expression levels to log-ratios. Unlike the correlation-based distance functions, the Euclidean distance takes the magnitude of the expression data into account. It, therefore, preserves more information about the data and may be preferable. De Hoon et al. (2002) used the Euclidean distance for k-means clustering.

| 2. | Normalized Euclidean distance | $d = \dfrac{d_E}{\sqrt{n}}$ |

where d_E is the Euclidean distance.

| 3. | Harmonically summed Euclidean distance | $d = \left[\dfrac{1}{n} \sum_{i=1}^{n} \left(\dfrac{1}{x_i - y_i} \right)^2 \right]^{-1}$ |

It is a variation of the Euclidean distance, where the terms for the different dimensions are summed inversely (similar to the harmonic mean), and is more robust against outliers compared to the Euclidean distance. Note that it is *not* a metric as it is not based on triangle inequality. For example, consider $u = (1, 0)$; $v = (0, 1)$; $w = (1, 1)$. This yields $d(u, v) = 1$ while $d(u, w) + d(w, v) = 0$

| 4. | City-block (or Manhattan) distance | $d = \sum_{i=1}^{n} |x_i - y_i|$ |

Alternatively known as the Manhattan distance or taxi cab distance, it is closely related to the Euclidean distance. Whereas the Euclidean distance corresponds to the length of the shortest path between two points, the city-block distance is the sum of distances along each dimension.

This is equal to the distance a traveller would have to walk between two points in a city. The city-block distance is a metric, as it satisfies the triangle inequality. As for the Euclidean distance, the expression data are subtracted directly from each other, and therefore should be made sure that they are properly normalized.

(Contd.)

(Table 3 Contd.)

Sl. No.	Distance or similarity measure	Expression Remarks
5.	Normalized City-block distance	$d = \dfrac{1}{n}\sum_{i=1}^{n}\left|x_i - y_i\right|$ or $d = \dfrac{d_C}{n}$ where d_C is the city block distance.
6.	Canberra distance	$d = \dfrac{1}{n}\sum_{i=1}^{n}\dfrac{\left|x_i - y_i\right|}{(x_i + y_i)},$ Result always falls in the range [0, 1].
7.	Bray-Curtis distance	$d = \dfrac{\sum\left|x_i - y_i\right|}{\sum(x_i + y_i)},$ Result always falls in the range [0, 1].
8.	Maximum coordinate difference	$d_{max} = \max\left|x_i - y_i\right|$ Computes maximum absolute difference along a coordinate.
9.	Minimum coordinate difference	$d_{min} = \min\left|x_i - y_i\right|$ Computes the minimum absolute difference along a coordinate.
10.	Dot product	$d_0 = -xoy$, where $xoy = \sum_{i} x_i \times y_i$ This is the dissimilarity version of the dot product measure.
11.	Pearson's correlation coefficient	$r = \dfrac{1}{n}\sum_{i=1}^{n}\left(\dfrac{x_i - \bar{x}}{\sigma_x}\right)\left(\dfrac{y_i - \bar{y}}{\sigma_y}\right)$ The Pearson distance is then defined as $d_p \equiv 1 - r$. In which $\bar{x}$ and $\bar{y}$ are the sample mean of x and y respectively, while σ_x and σ_y are the sample standard deviation of x and y. It is a measure for how well a straight line can be fitted to a scatter plot of x and y. If all the points in the scatter plot lie on a straight line, the Pearson correlation coefficient is either $+1$ or -1, depending on whether the slope of line is positive or negative. If it is equal to zero, there is no correlation between x and y. As the Pearson correlation coefficient fall between $[-1, 1]$, the Pearson distance lies between $[0, 2]$.
12.	Absolute Pearson's correlation	By taking the absolute value of the Pearson correlation, a number between [0, 1] is obtained. If the absolute value is 1, all the points in the scatter plot lie on a straight line with either a positive or a negative slope. If the absolute value is equal to 0, there is no correlation between x and y. The distance is defined as $d_A \equiv 1 - \left|r\right|$ where r *is* the Pearson correlation coefficient. As the absolute value of

(Contd.)

(Table 3 Contd.)

the Pearson correlation coefficient falls in the range [0, 1], the corresponding distance falls between [0, 1] as well. In the context of gene expression experiments, the absolute correlation is equal to 1 if the gene expression data of two genes/microarrays have a shape that is either exactly the same or exactly opposite. Therefore, absolute correlation coefficient should be used with care.

13. *Uncentred Pearson's correlation*

$$r_U = \frac{1}{n}\sum_{i=1}^{n}(\frac{x_i}{\sigma_x^{(0)}})(\frac{y_i}{\sigma_y^{(0)}}),$$

where $\sigma_x^{(0)} = \sqrt{\frac{1}{n}\sum_{i=1}^{n}x_i^2}$; and $\sigma_y^{(0)} = \sqrt{\frac{1}{n}\sum_{i=1}^{n}y_i^2}$.

The distance corresponding to the uncentred correlation coefficient is defined as $d_U \equiv 1-r_U$ where r_u is the uncentred correlation.

This is the same as for regular Pearson correlation coefficient, except that sample means $\bar{x}, \bar{y}$ are set equal to 0. The uncentred correlation may be appropriate if there is a zero reference state. For instance, in the case of gene expression data given in terms of log-ratios, a log-ratio equal to 0 corresponds to green and red signal being equal, which means that the experimental manipulation did not affect the gene expression. As the uncentred correlation coefficient lies in the range [–1, 1], the corresponding distance falls between [0, 2].

14. *Absolute uncentred Pearson's correlation*

$$d_{AU} \equiv 1-|r_U|$$

where r_u is uncentred correlation coefficient. As the absolute value of r_u falls between [0, 1], the corresponding distance also falls between [0, 1].

15. *Pearson's linear dissimilarity*

$$d_\rho(x,y) = \frac{1-\rho(x,y)}{2}, \text{ where } \rho(x,y) = \frac{(x-\bar{x})o(y-\bar{y})}{\sigma_x \times \sigma_y}$$

This is the dissimilarity version of the Pearson linear correlation p between two vectors. $D_\rho(i, j) \in [0, 1]$ with 0 indicating perfect similarity and 1 indicating maximum dissimilarity. Notice that p(i, j) is a type of normalized dot product.

16. *Pearson's absolute value dissimilarity*

$$d = \sqrt{\frac{n}{n-1}(d_{NormEuc}(x,y)^2 - q(x,y)^2)}$$

where $q(x,y) = \frac{1}{n}\left(\sum_i x_i - \sum_i y_i\right)$

$d_{NormEuc}$ is the normalized Euclidean distance.

(Contd.)

(*Table 3 Contd.*)

Sl. No.	Distance or similarity measure	Expression	Remarks
17.	Spearman's rank correlation		As in the case of the Pearson correlation, a distance measure corresponding to the Spearman rank correlation can be defined as $d_s \equiv 1 - r_S$ where r_s is the Spearman rank correlation. The Spearman rank correlation is an example of a non-parametric similarity measure. It is useful because it is more robust against outliers than the Pearson correlation. To calculate the Spearman rank correlation, each data value is replaced by their rank if the data in each vector is ordered by their value. Then the Pearson correlation between the two rank vectors instead of the data vectors is calculated. Weights cannot be suitably applied to the data if the Spearman rank correlation is used, especially since the weights are not necessarily integers.
18.	Kendall's τ distance	$$d = \sum_{i=1}^{n}(x_i - y_i)$$	A distance measure corresponding to Kendall's τ can be defined as Kendall's τ is another example of a non-parametric similarity measure. It is similar to the Spearman rank correlation, but instead of the rank those only, the relative ranks are used to calculate τ. As in the case of the Spearman rank correlation, the weights are ignored in the calculation. As Kendall's τ is defined such that it will lie between $[-1, 1]$, the corresponding distance will be in the range $[0, 2]$.
19.	Cosine distance	$$d = \frac{\sum\limits_{i=1}^{n} x_i y_i}{\sum\limits_{i=1}^{n} x_i \sum\limits_{i=1}^{n} y_i}$$	

DATA VISUALIZATION TECHNIQUES

A study of all available visualization techniques and their importance in analysis of gene expression data was carried out. It was found that different algorithms used different visualization techniques for analysis of the data. Some of the classical references are Caron et al. (2001), Chen et al. (2004), Eisen et. al. (1998), Ewing et. al. (2001), Kapushesky et al. (2004), Luo et al. (2003), Tavazoie et al. (1999), Taronen et al. (1999), Colantouni et al. (2000), de Hoon et al. (2004), Kohonen et al. (1996), Stanford Biomedical Informatics (2004), GEDA, etc.

Earlier, there were cases when the same visualization technique was used differently, such as the SOM output as seen in the difference in implementation of the GEPAS server, and the initial SOM version developed by Kohonen et al., 1996. While the former brought out clear and unique clusters as hexagonal grid, the latter produced many null clusters as well as similarly behaving clusters, which required altogether different visualization containing special colour coding and mapping scheme to generate heat map for proper identification. This helped in identification of similar clusters.

Kaski (1997) used SOM on various kinds of datasets, also emphasized on visualization through SOM and brought a comparison of k-means and SOM. Use of non-linear projection methods such as multi-dimensional scaling (MDS), Sammon's mapping, principal curves (which are generalization of principal components), triangulation method and replicator neural network were also discussed in the work.

The software should provide a platform for holding together all these visualization techniques for analysis of preprocessed data and processed data clusters through the use of group plots and individual plots. Facility should be provided to easily switch from one output form to another by selecting the corresponding button. Flexibility to save the images of visualization can also be provided so that they could be incorporated into the reports. For instance, histograms of all samples can be viewed in a single shot so as to ensure that they follow the normal distribution, as shown in the profile plots in Fig. 13.

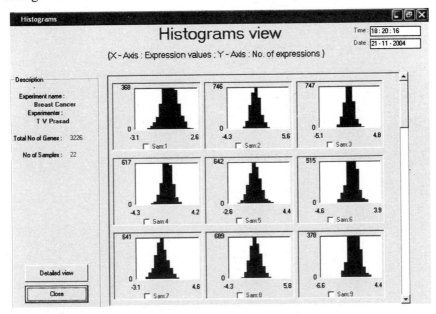

Fig. 13: Profile plots. One histogram for each sample—plots indicate that all samples follow normal distribution. The overall frequency vs. gene expression plot of the entire dataset matrix should also produce normal distribution.

It was observed that while one software suite presented well in terms of statistical computation, the other presented it in the form of equivalent graphs,

Table 4: Description of various visualization techniques for microarray gene expression data

Sl. No.	Visualization technique	Description or interpretation	Special considerations or features	Advantages and drawbacks	Complexity	Application	Figure	References
1.	Cluster view - (a) textual view	Clusters contain actual gene names, as generated in most of the text-based ANN software It is the most primitive of all and could be confusing for extremely large datasets.	None	Output is impressive, but could be difficult to understand. It does not give an idea of overall gene expression.	$O(n)$ if the entire matrix has been sorted on cluster number; otherwise $O(n^2)$.	SOM, LVQ, SVM and k-means; extended to HC and PCA.	15	GEPAS website of CNIO, Expression Profiler, GEDAS software, SOM-PAK and LVQ-PAK by Kohonen
2.	Cluster view - (b) temporal or wave graph	Clustered data is visualized in the form of a set of waves, in which each wave corresponds to a gene across samples on the X-axis. Also known as the temporal or wave graph view, this visualization technique can also be displayed as pie graph.	An extra wave is plotted in black colour to indicate average value of each sample in the cluster, which gives a fair idea of the expression level. It is also essential to display zoomed view of each cluster to enable scientists to see the expression behaviour in enlarged form.	Combined plot of all the genes can determine level of expression, overall as well as for specific genes. Very helpful for representing time series data. Requires a further GUI support to extract corresponding gene names. For a dataset containing discrete data (such as those containing numerous parametric values), this representation could render no meaningful use.	$O(n)$ if the entire matrix has been sorted on cluster number; otherwise $O(n^2)$.	SOM, LVQ, SVM and k-means; extended to HC and PCA	16, 17	GEPAS website of CNIO, Expression Profiler, GEDAS software, SOM-PAK and LVQ-PAK by Kohonen
3.	Heat map	Introduced as one of the most elegant graphical representations of cluster contents, it is very helpful in large scale data mining applications.	It can be further enhanced by use of colouring schemes to represent cluster of similar clusters. The colouring scheme conveys "logical classes" in the dataset or "knowledge" about	When number of clusters is too large, there can be many null or empty clusters, which convey no meaning and thus be eliminated. It can be applied to datasets having a number of features in them. Coloured sections or groups	$O(n^2)$	SOM, LVQ and k-means, SVM and HC	18	SOM-PAK LVQ-PAK and Kohonen by

(Contd.)

(Table 4 Contd.)

		certain hidden parameter of common amongst all the clusters.	of clusters give an idea of number of possible classes in the dataset. Requires data to be submitted in a specific format. When there are extremely large numbers of features (microarray gene expression does not have many), output could become cumbersome.			
4.	Dendrograph view	Also called as checks view, it is very similar to dendrogram output generated by Chen et al., 2000 or other graph visualization software. It can be used for visual inspection of raw, preprocessed and clustered data. This representation alone is not a true dendrogram output as it does not accompany gene tree and array tree.	User can control selection of colours for representing low to high gene expression such as green to red or blue to red or so on. Different colour codes can be assigned to represent null values or zero values. Shades represent intensity or magnitude of expression.	Most effective form of visualizing trend of gene expression in many samples and genes in one shot. However, if the dataset is very large, it requires another GUI support to extract the gene names. Very helpful for studying trend in time series data and data of same parameter over different samples.	O(n) if the entire matrix has been sorted on cluster number; otherwise O(n²).	19 SOM, LVQ and k-means, SVM and HC — Caron et al. (2001), GEDAS
5.	Microarray view	Very much similar to dendrograph view (or checks view) and is used for visual inspection of raw, preprocessed and clustered data.	Same as dendrograph view.		O(n²). Biggest drawback is that it requires sorting of the entire data matrix.	20 Biggest SOM, LVQ, k-means, HC, SVM and PCA — GEDAS software

(Contd.)

(*Table 4 Contd.*)

Sl. No.	Visualization technique	Description or interpretation	Special considerations or features	Advantages and drawbacks	Complexity	Application	Figure	References
6.	Proximity or distance map	It is a plot of distances between genes vs. genes similar to the distances table of various cities in the world as seen in the diaries. The gene expression matrix is sorted on cluster number, and then distance matrix is developed, which is a diagonal matrix. Each value is then displayed in the form of a coloured box. While white colour represents zero distance, black represents maximum distance. Diagonal line is always white indicating zero distance between same genes.	Black and white proximity map can be given a coloured effect by displaying all bands of genes in a single colour shade within a cluster.	With just a small plot, a fair view of cluster distances can be determined. It can provide the entire gene expression data matrix.	$O(n^2)$. The biggest drawback is that it requires sorting of the entire gene expression matrix.	SOM, LVQ, k-means, HC, SVM and PCA	21, 22,	GEDAS, GAP by Chen (2004)
7.	Whole microarray graph view	A highly versatile representation of gene expression data with each band (or line graph) corresponding to each sample. The portion between two horizontal lines contains expression values of 100 genes.	The last band is used to represent median of all the samples. Facility to change background colour and colour of bands provided. Visualization can be zoomed or reduced as per need.	While behaviour of all individual samples can be visually matched with neighbouring samples, it requires more time for representing more samples (when zoomed out).	$O(n^3)$	Raw data and preprocessed data; extended to SOM, LVQ, k-means, HC, SVM and PCA	23, 24,	GEDAS software
8.	Scatter plot	In addition to plot genes after PCA, it can be used for MA plots, preprocessing, etc. For PCA display of two PCs, one vertical and other horizontal. All data points are projected on these two lines.	Different colours for samples may be considered.	For PCA, output is processed for second time to project all data points on the PCs.	$O(n)$	PCA	25	Cluster software by de Hoon et al. (2004), GEDAS, GEDA software

(*Contd.*)

(*Table 4 Contd.*)

No.	View	Description	Features	Comments	Complexity	Applies to	Ref.	Software
9.	Principal component view	PC view is a line graph drawn as sum of principal components (Eigen value) and individual expression values. Though all components are displayed, the first two or three PCs play important role in dimensionality reduction.	A good PC line graph always falls down on X axis exponentially. Facility provided to change colours of PCs. Graph can be redrawn with required number of PCs.	First two or three PCs are usually sufficient to generate the entire dataset.	$O(n)$	PCA	26	GEDAS, Cluster software by de Hoon et al. (2004)
10.	Histogram	It is a common form of visualizing basic statistical data.	None	Cannot be used for complex data forms.	$O(n)$	All	27	None
11.	Tree or dendrogram view	One of the most effective and powerful representations of clustered gene expression data, consisting of three portions viz., gene tree, array tree and colour coded band of gene expression. Also known as matrix tree plot or 2-way dendrograms. In HC, it is the most common output. In order to standardize and provide common outputs to all data mining applications, output was converted into cluster view that further led to views such as microarray, textual, whole genome, etc.	Same as dendrograph view	Offers clustering of both genes and samples simultaneously. However, if the dataset is very large, it also requires another GUI support to extract the gene names. Very helpful for studying the trend in time series data and data of same parameter over different samples. Inter-date relationship will be lost by representing multi dimensional data in a 2D tree format.	$O(n^3)$; two different algorithms work together to build up the gene tree on one side, and array tree and the colour coded band of expression values on the other.	HC	28	TreeView software by Eisen et al. (1998), GEDAS, Luo et al. (2003)

(*Contd.*)

(Table 4 Contd.)

Sl. No.	Visualization technique	Description or interpretation	Special considerations or features	Advantages and drawbacks	Complexity	Application	Figure	References
12.	Decision-space or search-space view	Used for classification, either SVM or LVQ, and this kind of representation of gene expression data come very handy with each decision space corresponding to each class of genes. The figure was extracted from a demo version of SVM, it could be applied to any multi-class classification.	Coloured decision spaces give very impressive look and better understanding of the data grouping.	Requires data to be in 2D form; higher dimensional representation could be very complex.	Not available	LVQ and SVM. Could be extended to SOM, k-means, HC, PCA, etc. if it is possible to represent their output in 2D.	29	LibSVM
13.	Tree-map view	Applied to the results of gene expression data from hierarchical clustering. There are a large number of Tree-map visualization variants for representing hierarchical relationships.	Colour and depth variation could be effectively used to form clusters, which further exhibit information such as cluster size, overall expression, etc.	Visually very attractive for smaller datasets; can be very confusing or scrammed for larger datasets.	Not available	HC, could be extended to other clustering and/or classification techniques	30	TreeView of Eisen et al. (1998), GEDAS, Expression Profiler, Cleaver 1.0
14.	Box-Whisker plot	Very handy in dealing with raw and preprocessed gene expression data. The plot provides information on overall over and under expression along with mean, upper and lower quartile together in one plot. It can, in many cases, be applied to reduced or transformed datasets with large number of insignificant genes and samples pruned.	Colour variation could be introduced for different samples, upper and lower quartile variation.	Mean, median as well as upper and lower quartile can be viewed simultaneously; useful for preliminary analysis and when a number of genes and/or samples have been eliminated, causing change in the mean and other parameters.	$O(n^2)$	Raw data and preprocessed data	31	GEDAS, GEDA
15.	Gene Ontology	There can be different visualization technique to see the processed information in a single shot.	None	Depending on the requirements, the visualization can be planned.	$O(n)$	All	32	All software

and still further another suite presented the output in altogether different graphical form. Graphs such as group plots, MA plots, etc. that have great significance in biological data mining (Fig. 14). Details of each visualization technique have been described in Table 4.

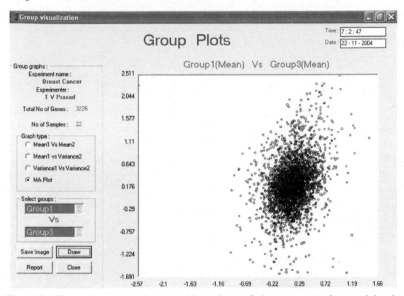

Fig. 14: The MA group plot provides view of the amount of spread in the overall gene expression pattern in the data groups. These plots could be obtained for data before transformation and cannot be obtained for negative or zero data.

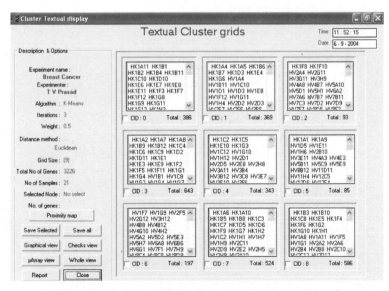

Fig. 15: The textual view exhibits the same clusters as that of temporal view; however, actual gene names are displayed. In this example, a 3 × 3 SOM configuration output can be seen.

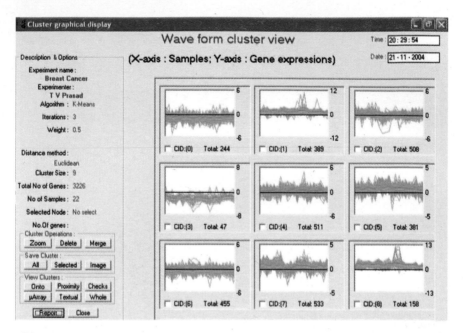

Fig. 16: The temporal/wave graph view. Each cluster contains a number of genes represented as a wave, the green line is the average expression level.

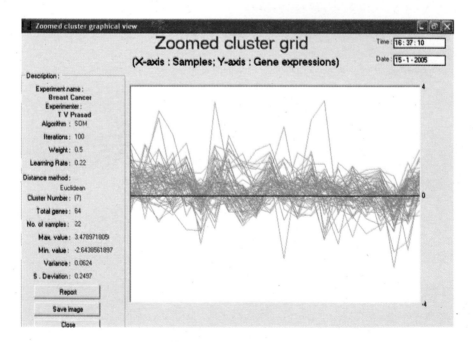

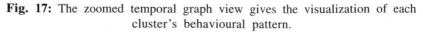

Fig. 17: The zoomed temporal graph view gives the visualization of each cluster's behavioural pattern.

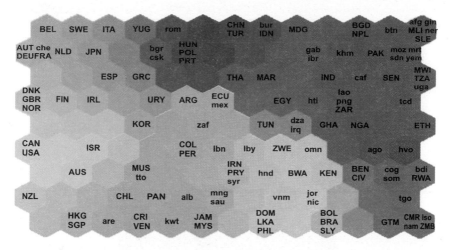

Fig. 18: Coloured heat map showing clusters of similarly behaving entities as group of cells in same colour/shade.

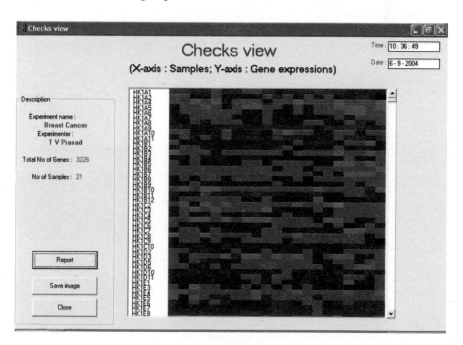

Fig. 19: The checks view (or dendrograph view) visualization is a well-known method of plotting the gene expression levels through colour coded boxes. It gives a fair idea of gene expression and can be applied to time series as well as discrete data. It has been applied to raw, preprocessed and clustered data.

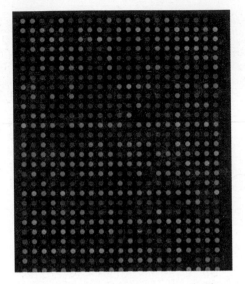

Fig. 20: The microarray view is very similar to the dendrograph or checks view, except that the individual genes are represented in circular spots. It is a good visualization for having visual inspection of gene expression data.

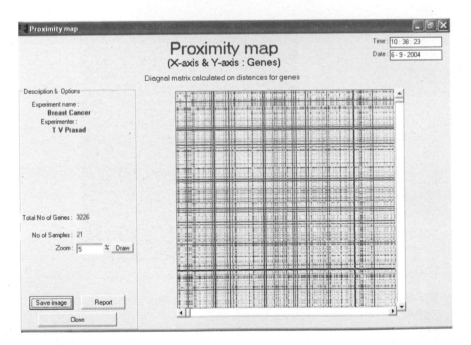

Fig. 21: The proximity map visualization is a map of distances plotted genes vs. genes. Usually, it is symmetric/diagonal map. It has been applied to raw, preprocessed and clustered data.

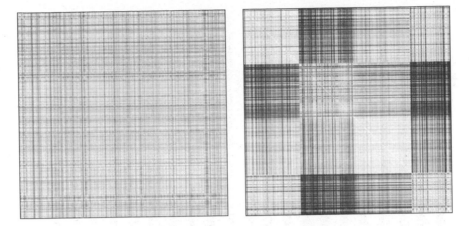

Fig. 22: The proximity map visualization is a map of distances plotted genes vs. genes. Usually, it is symmetric/diagonal map. Proximity maps of raw data set and clustered data can be seen here.

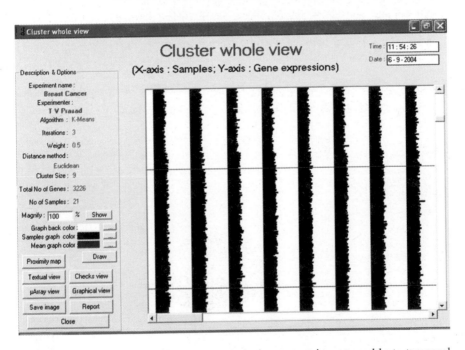

Fig. 23: The whole (microarray or genome) view output is comparable to temporal view. The samples are always maintained; however, the arrangement of genes may alter after clustering. It is the best suited visualization for having visual inspection of gene expression.

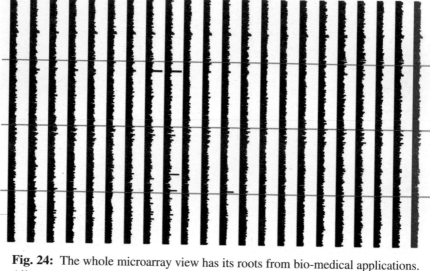

Fig. 24: The whole microarray view has its roots from bio-medical applications. All samples and genes can be visualized in such a form that dis(similarity) can be easily obtained. The last strip/band is the median of all samples. Some genes falling between 201 and 300 in samples 8 and 9 have very high gene expression; similarly in sample 11 between genes 301 and 400.

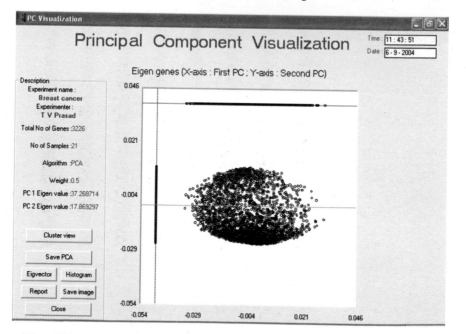

Fig. 25: The principal component view is essentially a scatter graph in a way that the space of the genes/samples plotted is usually elliptical.

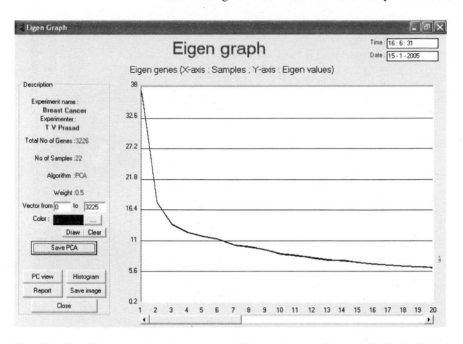

Fig. 26: The Eigen graph is in no way different from a line graph. Each Eigen line is the sum of principal component and individual gene expression value.

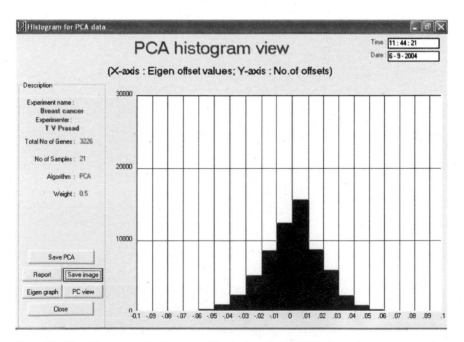

Fig. 27: The histogram is the most fundamental visualization of gene expression data and can be applied to all forms of datasets.

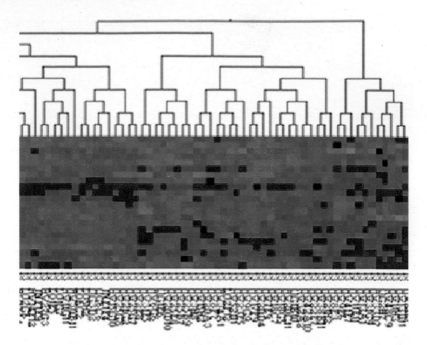

Fig. 28: The dendrogram or tree view can be constructed for genes as well as samples.

Fig. 29: The decision space view gives an idea of the number of classes (or groups) that can be made from the given data.

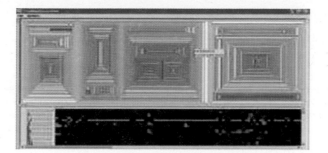

Fig. 30: A different tree-map visualization.

Fig. 31: The Box-Whisker plot provides most of the statistics in one graph covering maximum, minimum, median, upper quartile and the lower quartile.

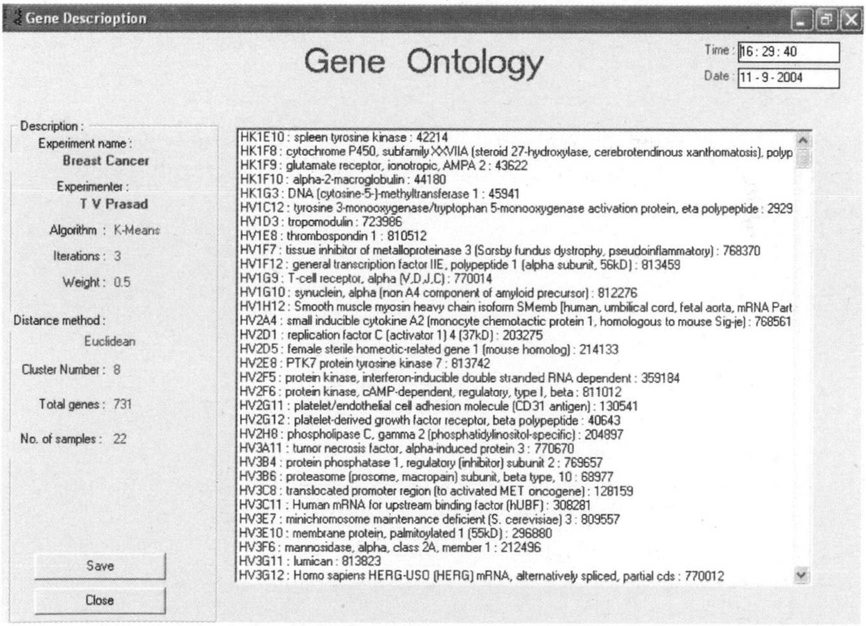

Fig. 32: The gene ontology view is not a true visualization technique, but a listing of gene number together with their details like gene names and function, if available in the dataset. It is helpful to further access or understanding the function of genes in a particular cluster.

Features Table

A consolidation of the views that can be provided for visualization vs. algorithms has been listed in Table 6.

Table 6: Applicability of visualization techniques in various contexts

Visualization/ Algorithm	Raw data	Pre-processed data	SOM	k-Means	LVQ	HC	PCA (gene)	SVM
Histogram	✓	✓					✓	
Checks view	✓	✓	✓	✓	✓	✓	✓	✓
Microarray	✓	✓	✓	✓	✓	✓	✓	✓
Whole sample	✓	✓	✓	✓	✓	✓	✓	✓
Proximity map	✓	✓	✓	✓	✓	✓	✓	✓
Temporal (incl. zoomed cluster view)			✓	✓	✓	✓	✓	✓
Textual			✓	✓	✓	✓	✓	✓
PC view							✓	
Eigen graph							✓	
Tree view						✓		
Scatter plot							✓	
Decision (or search) space					✓			✓
Box-Whisker plot	✓	✓						
Gene Ontology			✓	✓	✓	✓	✓	✓

A variety of other visualizations for microarray gene expression data are available. The websites of GEDA and SilicoCyte can be converted or presented using the techniques as discussed above:

- 2D and 3D score plots
- Profile plots
- Scatter plots
  - 3D scatter plots
  - PCA visualization - Result on 3D scatter plot
  - Isomap visualization
  - Multi-dimensional scaling
- Venn Analysis Diagrams for visualizing similar elements in microarrays, same genes in gene groups or same genes in microarray designs
- SOM visualization of clustering result on
  - 2D Trellis plot
  - U-matrix
  - 1D SOM

There are a number of visualization techniques to represent interaction between the genes in a genetic regulatory network. These network types have been standardized (Hwang et al., 2001).

SORTING OF DATASET BEFORE CLUSTERING

Sorting of data matrix based on the relative distances computed from one to another can result in better grouping of the genes. Distance from the first gene to all other genes can be computed and arranged in ascending order, then distance from the second gene to the remaining genes computing and arranged in ascending order, and the process continued till all the genes are placed in order. Entire dataset can be then clustered using the SOM algorithm after which the placement of genes gets enhanced considerably. The quality of visualization also gets better. The resultant two proximity maps, as shown in Fig. 33, indicate that though the data is clustered, the cluster distances became much clearer and the genes are presented in sorted order.

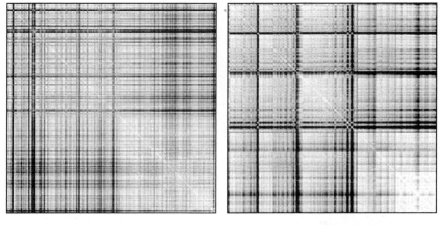

(a) Proximity map of SOM output	(b) Proximity map of SOM output
(without sorting)	after sorting data

Fig. 33: Enhancing the proximity map by sorting the data before clustering.

For obtaining phenotype classification, the ANN clustering algorithms can be the best compared to other non-ANN based techniques, (Stolovitzky et al., 2004). It is essential to cluster both the genes and the samples for obtaining the best phenotypes. The phenotypic extraction reveal:

- If only similar genes are extracted - they are more likely functionally related genes.
- If only similar samples are extracted - they are similar or closely related tissue, organ, organism, or disease.
- Subset of both genes and samples are similar - it is called a phenotype.
- Contrarily, if subset of both genes and samples are widely varying - this should not be omitted but can provide the best meaning as to why the genes that expressed high in a particular group were low expressing in the other. Especially, when one of the groups is control, then the purpose would have fulfilled.

A step further, clustering of samples based on the sorted matrix obtained from such experiment help in presenting the data matrix sorted on both genes (within clusters) as well as samples.

DISCUSSIONS

Pre-processing

In most of the cases, the ANN models demand considerable amount of pre-processing before actually going for cluster analysis. Certain pre-processing techniques such as discretization of data, internal sorting, etc. can improve the computational speed of the algorithms. Removal of genes that fall under a low lying band such as between the expression level of -1 and 1, which is also the band for noise-like features, can be eliminated from further computation. A combination of these techniques could also prove to be highly useful in cluster analysis.

There has been a continuous debate over whether removal of genes would be proper or not. Though, no concrete decision has been made, it is felt that pre-processing can be done as far as study remains focussed on identifying functions of certain group of genes; however, the impact of such pruning or manipulation of genetic data for the purpose of drug designing could be strong.

Visualization

Unlike SOM and LVQ models, most of the models like perceptron and other feed-forward neural networks are not suitable for visualization. Through GEDAS a host of visualization techniques were introduced and all algorithms were standardized so that same set of outputs could be used. The advantage of doing so brought a clear comparison between applications of different algorithms.

All software available publicly are either too much tilted towards computational aspects or only towards visualization. Moreover, hardly any software tool provides a common platform for analysis/modeling of various gene expression datasets. The TreeView software by Eisen Lab was reviewed to incorporate changes to fit duplicates and to generate the output from the raw input data instead of the distance matrix. This has brought considerable enhancement to the visualization as both the clarity in tree generation and the checks (or heat map) was noticeable.

Application of Various Clustering Techniques such as SOM, HC, k-means, PCA

The clustering of gene expression data using various clustering techniques results in different outcomes. Whereas cluster analysis using SOM is well known, the combination of PCA for dimensionality reduction before application of SOM brings out that the combination theory would work only when the first algorithm is for pattern classification. Since LVQ (through the three variants) does the actual grouping of genes, it does not bring any noticeable change on application of PCA before it.

Clustering using PCA can be highly consistent and logical. The formation of clusters, though only two (due to bifurcation), provides clusters with considerable amount of relationship amongst the genes in clusters.

It is worth noting that all the methods use the conventional Euclidean distance for computing misclassification. It is worth pondering as to why not use any other distance measure such as Pearson correlation coefficient or Bray-Curtis distance measure for the purpose.

Data Mining

The definitions of clustering and classification overlap largely, whereas the idea behind both is common - to group or categorize the data into homogenous sets or groups. The standardization of hierarchical clustering output with other clustering techniques vanishes the difference. The standardization of other output forms in GEDAS homogenized the purposes of clustering and classification.

The gene expression data alone do not convey the complete information at any given point of time. Instances and information such as hereditary or history of cancer cases in the family, exposure to strong chemicals, food habits, drugs being used for cure/prevention, surgeries or therapies undertaken, etc. do not form part of the experimental process even during the data analysis stage. It is quite possible that unwanted variation in any one of these parameters/observations may produce faulty or highly varying results.

Data Preprocessing and Extraction

While working with the hierarchical clustering of gene expression data, it may be noticed that Pearson's correlation should be preferred as distance measure, even though Euclidean and Manhattan distances provide excellent clustering analysis, since the purpose of using HC is to build a correlation matrix rather than a distance matrix. An additional parameter "Tree depth" to the tree output can be very useful through the implementation, which can be used for conversion of Tree View (dendrogram) output into cluster view format. Also, the outputs of average- and complete-linkage methods are found to be better than the single linkage method.

Using the k-means clustering, no two experiments yield identical results. This leads to the fact that the random seed involved in selection of cluster centroid yield clusters of varying sizes, and hence, the results remain untraceable.

It is purely the prerogative of the user to utilize these algorithms/models for extracting different outputs, which can be any of the following order:

- Similarly or differentially expressing
- High, averagely or low expressing
- Across genes or samples
- Obtaining phenotypes or detection of consistently varying gene expression levels, i.e. high in some but low in others, or vice versa.

PROBLEMS AND SUGGESTIONS FOR FUTURE WORK

1. Compare and contrast different standards of microarray technology currently available.
2. Overall, there is a significant requirement of standardization in the microarray gene expression data analysis scenario, which currently suffers from the following limitations due to non-standardization, and consequently, it can be concluded that efforts would be needed to formalize on which platform, algorithm, technique/model, parameter, and visualization to be used for analysis of various kinds of data:
   - Microarray production and techniques not standardized
   - Preprocessing techniques not yet standardized
   - No standardized process or model available for analysis
   - No standardized visualization techniques available
   - Standardization of minimal information exchange on microarray data brought out recently under the MIAME project

   Support for the text, Microsoft Excel and Microsoft Access based datasets, extension to read and process datasets stored in other formats such as Oracle, Sybase, and numerous other databases can be provided. Addition of suitable parameters such as function of the genes, class to which they belong, length, etc. parameters to the microarray gene expression data matrix will help mining the data in a much precise manner than being done conventionally.
3. Clustering of gene expression data: The work can be extended to include the following models, both ANN-based, and non ANN based:
   - Classification algorithms like the Linde-Buzo-Gray (LBG) or the Generalized Lloyd algorithm, etc. could be incorporated (Phani Kumar, 2004). Training the ANNs using the LBG can further improve the testing capability of LVQ, SVM as well as other clustering techniques considerably. Mutual nearest neighbours (Jarvis-Patrick) clustering method is considered to be a robust technique and is particularly useful on very large data sets.
   - Other clustering algorithms such as fuzzy c-means, improved k-means, self-organizing tree algorithm, multi-layer perceptron, genetic algorithm, agglomerative algorithms, etc. can also be integrated into the software suite.

4. Other distance measures such as unnormalized symmetric Kullback-Liebler measure, weighted dot product, Spearman's rank dissimilarity, etc. can be implemented to provide more flexibility to the user. Some measures such as the unnormalized symmetric Kullback-Liebler measure do not require normalization (or sometimes even pre-processing) of data. Testing of various clustering and classification algorithms using these distance measures can be a challenging work.

5. Better visualization for better data mining: Better user interfaces such as facility to select a gene or group of genes from a cluster on any visualization technique, representation of clusters on the proximity map in different colours, better organization of experiments, automation and auto-comparison of experiments, etc. could be added to the software suite to enhance the power of both visualization and data mining.

6. Better interaction with SVM can be developed, as presently, the outputs of clustering techniques cannot be provided as input to the SVM. Multi-class classification using SVM can also be carried out. While doing so, the visualization should also be developed as the one provided by Chang and Lin (2004) in the SVMToy in which different classes can be identified with the search (or decision) space in different colours.

7. Three dimensional visualization/representation of output from the PCA and other clustering algorithms can be developed, which will give a better understanding of the datasets and the similarity of genes/samples, within and outside a given population.

8. The output visualization of Tree View (dendrogram) visualization requires further review to incorporate additional features. A host of newer visualization techniques, to support the data mining algorithms can be incorporated into the software suite, which may include, but not limited to, the following (Eisen et al., 1998):

- Matrix Tree Plots (1 and 2-way Dendrograms); recently, the dendrograms are also available in 3D form.
- Scatter plots could be standardized for visualization of all outputs using 2D and 3D scatter plots.
- Venn Analysis Diagrams for visualizing similar elements in microarrays, same genes in gene groups or same genes in microarray designs.
- For PCA visualization, though the results were presented successfully on 2D scatter plot, representation using 3D scatter plot could be more effective.
- Tree-map visualization of various clusters in different clusters could prove to be one of the best techniques for providing web interface to any microarray gene expression data analysis software, with retrieval of gene ontology on the basis of mouse click on a particular cluster.
- The SOM output was presented through a 1D/2D visualization; however, 3D form could be added to provide a better understanding of the processed results.

REFERENCES

Articles/Papers/Presentations/Books

Al-Kanhal, M.I. and Al-Hendi, R.I. (1992). Arabic phoneme map based on vector quantization neural networks. Graduate Thesis, King Saud University, Saudi Arabia.

Alizadeh, A.A., Eisen, M.B., Davis, R.E., Ma, C., Lossos, I.S., Rosenwald, A., Boldrick, J.C., Sabet, H., Tran, T., Yu, X., Powell, J.I., Yang, L., Marti, G.E., Moore, T., Hudson, J. Jr., Lu, L., Lewis, D.B., Tibshirani, R., Sherlock, G., Chan, W.C., Greiner, T.C., Weisenburger, D.D., Armitage, J.O., Warnke, R., Levy, R., Wilson, W., Grever, M.R., Byrd, J.C., Botstein, D., Brown, P.O. and Staudt, L.M. (2000). Distinct types of diffuse large B-cell lymphoma identified by gene expression profiling. *Nature*, **403(3):** 503-511.

Alter, O., Brown, P.O. and Botstein, D. (2000). Singular value decomposition for genome-wide expression data processing and modeling. *Proc. Natl. Acad. of Sc. USA*, **97(18):** 10101-10106.

Anderson, J.A. (2001). An Introduction to Artificial Neural Networks. Prentice Hall of India, New Delhi.

Baldi, P. and Brunak, S. (2003). Bioinformatics: The Machine Learning Approach. Affiliated East-West Press Pvt. Ltd., New Delhi.

Baldi, P. and Hatfield, G.W. (2001). Microarrays and Gene Expression. Cambridge University Press, Cambridge.

Bassett, D. Jr, Eisen, M.B. and Boguski, M.S. (1999). Gene Expression Informatics - it's all in your mind. *Nature Genetics*, Supplement **21**.

Ben-Dor, A., Shamir, R. and Yakhini, Z. (1999). Clustering gene expression patterns. *Journal of Computational Biology*, **6(3/4):** 281-297.

Bergeron, B. (2003). Bioinformatics Computing. Prentice Hall of India, New Delhi.

Bowtell, D. (1999). Options available - from start to finish - for obtaining expression data by microarray. *Nature Genetics*, Supplement **21**.

Brazma, A., Hingamp, P., Quackenbush, J., Sherlock, G., Spellman, P.T., Stoeckert, C., Aach, J., Ansorge, W., Ball, C.A., Causton, H.C., Gaasterland, T., Glenisson, P., Holstege, F.C.P., Kim, I.F., Markowitz, V., Matese, J.C., Parkinson, H., Robinson, A., Sarkans, U., Schulze-Kremer, S., Stewart, J., Taylor, R., Vilo, J. and Vingron, M. (2001). Minimum information about a microarray experiment (MIAME) - toward standards for microarray data. *Nature Genetics*, **29:** 365-371.

Caron, H., van Schaik, B., van der Mee, M., Baas, F., Riggins, G., van Sluis, P., Hermus, M.C., van Asperen, R., Boon, K., Voute, P.A., van Kampen, A. and Versteeg, R. (2001). The Human Transcriptome Map: Clustering of highly expressed genes in chromosomal domains. *Science*, **291:** 1289-1292.

Carr, D.B., Somogyi, R. and Micheals, G. (1997). Templates for looking at gene expression clustering. Stat. Comput. & Stat. Graph. *Newsletter*, 20-29.

Chakraborty, C. (2004). Bioinformatics: Approaches and Applications. Biotech Books, Delhi.

Chee, M.C., Yang, R., Hubbell, E., Berno, A., Huang, X.C., Stern, D., Winkler, J., Lockhart, D.J., Morris, M.S. and Fodor, S.P.A. (1996). Accessing genetic information with high-density DNA arrays. *Science*, **274:** 610-614.

Chen, D., Chang, R.F. and Huang, Y.L. (2000). Breast cancer diagnosis using self-organizing map for sonography. Ultrasound Medical Biology, **26(3):** 405-411.

Chen, C.H. et al. (2004). Generalized Association Plots (GAP), Presentation on "Cluster Analysis and Visualization". *In:* Workshop on Statistics and Machine Learning, Institute of Statistical Science.

Cho, S.B. and Won, H.H. (2003). Machine learning in DNA microarray analysis for cancer classification. Conferences in Research and Practice in Information Technology, 19 (Ed. Yi-Ping Phoebe Chen, Australian Computer Society).

Churchill, G.A. (2002). Fundamentals of experimental design for cDNA microarrays. *Nature Genetics*, 32 Suppl: 490-495.

D'haeseleer, P., Wen, X., Fuhrman, S. and Somogyi, R. (1997). Mining the gene expression matrix: Inferring gene relationships from large scale gene expression data. *In:* Information processing in cells and tissues (eds. Paton, R.C. and Holcombe, M.). Plenum Press, 203-212.

DeRisi, J., Penland, L., Brown, P.O., Bittner, M.L., Meltzer, P.S., Ray, M., Chen, Y., Su, Y.A. and Trent, J.M. (1996). Use of a cDNA microarray to analyze gene expression patterns in human cancer. *Nature Genetics*, 14(4): 457-460.

Dopazo, J. (2002). Microarray data processing and analysis. *In:* Microarray Data Analysis II. Kluwer Academic Publ., 43-63.

Dudoit, S. and Gentleman, R. (2002a). Cluster analysis in DNA microarray experiments. Bioconductor Short Course, Presentation slides.

Dudoit, S., Fridlyand, J. and Gentleman, R. (2002b). Classification analysis in DNA Microarray experiments. Bioconductor Short Course, Presentation slides.

Durbin, B.P., Hardin, J.S., Hawkins, D.M. and Rocke, D.M. (2002). A variance-stabilizing transformation for gene-expression microarray data. *Bioinformatics*, 18(90001): S105-S110.

Eijssen, L. (2000). Cluster analysis of microarray gene expression data. Master's thesis, Faculty of General Sciences, Maastricht University, The Netherlands.

Eisen, M.B., Spellman, P.T., Brown, P.O. and Botstein, D. (1998). Cluster analysis and display of genome-wide expression patterns. Proc. Natl. Acad. of Sc. USA, 95: 14863-14868.

Ewing, R.M. and Cherry, J.M. (2001). Visualization of expression clusters using Sammon's non-linear mapping. *Bioinformatics*, 17(7).

Freeman, J.A. and Skapura, D.M. (1991). Neural Networks. Addison Wesley, USA.

Furey, T.S., Cristianini, N., Duffy, N., Bednarski, D.W., Schummer, M. and Haussler, D. (2000). Support vector machine classification and validation of cancer tissue samples using microarray expression data. *Bioinformatics*, 16: 906-914.

Golub, T.R., Slonim, D.K., Tamayo, P., Huard, C., Gaasenbeek, M., Mesirov, J.P., Coller, H., Loh, M.L., Downing, J.R., Caligiuri, M.A., Bloomfield, C.D. and Lander, E.S. (1999). Molecular classification of cancer: Class discovery and class prediction by gene expression monitoring. *Science*, 286: 531-537.

Hacia, J.G., Brody, L.C., Chee, M.S., Fodor, S.P. and Collins, F.S. (1996). Detection of heterozygous mutations in BRCA1 using high density oligonucleotide arrays and two-colour fluorescence analysis. *Nature Genetics*, 14: 441-447.

Han, J. and Kamber, M. (2001), Data Mining: Concepts and Techniques. Elsevier, San Francisco, USA. .

Haykin, Simon (1999). Artificial Neural Networks: A Comprehensive Foundation 2nd ed. Addison Wesley.

Hedenfalk, I., Duggan, D., Chen, Y., Radmacher, M., Bittner, M., Simon, R. et al. (2001). Gene-expression profiles in hereditary breast cancer. *New England Journal of Medicine*, 344: 539-548.

Herroro, J., Valencin, A. and Dopazo, J. (2001). A hierarchical unsupervised growing neural network for clustering gene expression patterns. *Bioinformatics*, **17:** 126-136.

Hwang, K.B., Cho, D.Y., Park, S.W., Kim, S.D. and Zhang, B.T. (2001). Applying machine learning techniques to analysis of gene expression data: Cancer Diagnosis. *In:* Methods of Microarray Data Analysis. Kluwer Academic, 167-182.

Iyer, V.R., Eisen, M.B., Ross, D.T., Schuler, G., Moore, T., Lee, J.C.F., Trent, J.M., Staudt, L.M., Hudson Jr. J., Boguski, M.S., Lashkari, D., Shalon, D., Botstein, D. and Brown, P.O. (1999). The transcriptional program in response of human fibroblasts to serum. *Science*, **283:** 83-87.

Jagota, Arun (2001). Microarray data analysis and visualization. Dept. of Computer Engineering, University of California, CA., USA.

Kaski, S. (1997). Data exploration using self-organizing maps. Doctor of Technology Thesis, Helsinki University of Technology, Espoo, Finland.

Kapushesky, M., Kemmeren, P., Culhane, A. C., Durinck, S., Ihmels, J., Körner, C., Kull, M., Torrente, A., Sarkans, U., Vilo, J. and Brazma, A. (2004). Expression Profiler: next generation-an online platform for analysis of microarray data. *Nucleic Acids Research*, **32** (Web Server issue): W465-W470.

Khan, J., Wei, J.S., Ringnér, M., Saal, L.H., Ladanyi, M., Westermann, F., Berthold, F., Schwab, M., Antonescu, C.R., Peterson, C. and Meltzer, P.S. (2001). Classification and diagnostic prediction of cancers using gene expression profiling and artificial neural networks, *Nature Medicine*, **7(6):** 673-679.

Klingbiel, D. (2003). Singular value decomposition for feature selection in cDNA arrays, Talk at Max Plank Institute for Molecular Genetics, Germany, available at http:// compdiag.molgen.mpg.de/docs/talk_03_03_03_klingbiel.pdf.

Koren, Y. and Carmel, L. (2003). Visualization of labeled data using linear transformation, Proceedings of IEEE Information Visualization (InfoVis '03), IEEE, pp. 121-128, Presentation slides, available at http://www.cs.ubc.ca/~tmm/courses/ cpsc533c-04-spr/ slides/update.0317.mtan.ppt.

Kurimo, M. (1997). Using self-organizing maps and learning vector quantization for mixture density hidden Markov models. Doctor of Technology Thesis, Helsinki University of Technology, Espoo, Finland.

La Vigna, A. (1989). Non-parametric classification using learning vector quantization. Ph.D. thesis, University of Maryland, USA.

Li, L., Weinberg, C.R., Darden, T.A. and Pederson, L.G. (2001). Gene selection for sample classification based on gene expression data: Study of sensitivity to choice of parameters of the GA/KNN method. *Bioinformatics*, **17(12),** 1131-1142.

Liao, L. (2002). Clustering and classification and their applications in bioinformatics. Lecture notes, Discovery Information and High Performance Computing (ELEG 667).

Luo, F., Tang, K. and Khan, L. (2003). Hierarchical clustering of gene expression data. University of Dallas, TX, USA.

Mount, D.W. (2001). Bioinformatics: Sequence and Genome Analysis. Cold Spring Harbor Laboratory Press, NY, USA.

Narayanan, A., Keedwell, E.C. and Olsson, B. (2003). Artificial intelligence techniques for bioinformatics. *Applied Bioinformatics, Open Mind Journals*.

Nilsson, J. (2002). Methods for classification of gene expressions. Master's thesis, Centre for Mathematics, Lund University, Lund, Sweden.

Phanikumar, B. (2002). Clustering algorithms for microarray data mining. Masters' Thesis, Institute of Systems Research, University of Maryland, USA.

Pocock, M.R. and Hubbard, T.J.P. (2000). A browser for expression data. *Bioinformatics*, **16(4)**.

Prasad, T.V. and Ahson, S.I. (2005a). Visualization of microarray gene expression data. *Bioinformation*, **2006**.

Prasad, T.V. and Ahson, S.I. (2005b). Application of Learning Vector quantization on microarray gene expression data. *Bioinformation*, submitted.

Prasad, T.V., Ravindra Babu, P. and Ahson, S.I. (2005c). GEDAS - Gene Expression Data Analysis Suite Software, *Bioinformation*, 2006.

Quackenbush, J. (2002). Microarray data normalization and transformation. *Nature Genetics*, **32** Suppl: 496-501.

Ramaswamy, S., Tamayo, P., Rifkin, R., Mukherjee, S., Yeang, C.H., Angelo, M., Ladd, C., Reich, M., Latulippe, E., Mesirov, J.P., Poggio, T., Gerald, W., Loda, M., Lander, E.S. and Golub, T.R. (2001). Multiclass cancer diagnosis using tumor gene expression signatures. *Proc. Natl. Acad. of Sc.* USA, **98(26)**: 15149-15154.

Raychaudhuri, S., Stuart, J.M. and Altman, R.B. (2000). Principal components analysis to summarize microarray experiments: Application to sporulation time series. *Pacific Symposium of Biology*, **5**: 452-463.

Sharan, R., Elkon, R. and Shamir, R. (2001). Cluster analysis and its applications to gene expression data. Ernst Schering Workshop on Bioinformatics and Genome Analysis, Springer Verlag.

Sing, J.K., Basu, D.K., Nasipuri, M. and Kundu, M. (2003). Improved *k*-means algorithm in the design of RBG neural networks. *Proceedings of IEEE TENCON* 2003, Bangalore, India, October 2003.

Slonim, D., Tamayo, P., Mesirov, J., Golub, T.R. and Lander, E. (2000). Class prediction and discovery using gene expression data. Proceedings of RECOMB 2000.

Spellman, P.T., Miller, M., Stewart, J., Troup, C., Sarkans, U., Chervitz, S., Bernhart, D., Sherlock, G., Ball, C.A., Lepage, M., Swiatek, M., Marks, W.L., Goncalves, J., Market, S., Iordan, D., Shojatalab, M., Pizarro, A., White, J., Hubley, R., Deutsch, E., Senger, M., Aronow, B.J., Robinson, A., Bassett, D., Stoeckert, J. Jr. and Brazma, A. (2002). Design and implementation of microarray gene expression markup language (MAGE-ML). *Genome Biology*, **3(9)**.

Stolovitzky, G., Lepre, J. and Tu, Y. (2004). Gene expression pattern discovery in gene expression microarrays. Presentation slides, available at http://www.ibm.com/solutions/lifesciences.

Szallasi, Z. (1998). Gene expression patterns and cancer. *Nature Biotechnology*, **16**: 1292-1293.

Talavera, L. (2000). Dependency-Based Feature Selection for Clustering Symbolic Data. *Intelligent Data Analysis*, **4**: 19-28.

Tavazoie, S., Hughes, J.D., Campbell, M.J., Cho, R.J. and Church, G.M. (1999). Systematic determination of genetic network architecture. *Nature Genetics*, **22**: 281-285.

Tibshirani, R., Hastie, T., Eisen, M., Ross, D., Botstein, D. and Brown, P. (1999). Clustering methods for the analysis of DNA microarray data. Technical Report, Stanford University, USA.

Toronen, P., Kolehmainen, M., Wong, G. and Castren, E. (1999). Analysis of gene expression data using self-organizing maps. *FEBS Letters*, **451(2)**: 142-146.

Vijaya, P.A., Murty, M.N. and Subramaniam, D.K. (2003). An efficient incremental protein sequence clustering algorithm. Proceedings of IEEE TENCON 2003, Bangalore, India, October 2003.

Vipin Kumar (2002). Data Mining Algorithms. Tutorial at IPAM 2002, Presentation slides.

Wall, M.E., Rechtsteiner, A. and Rocha, L.M. (2003). Singular value decomposition and principal component analysis. In: A Practical Approach to Microarray Data Analysis (eds. Berrar, D.P., Dubitzky, W., Granzow, M.), 91-109, Kluwer, MA, USA.

Wen, X., Fuhrman, S., Michaels, G.S., Carr, D.B., Smith, S., Barker, J.L. and Somogyi, R. (1998). Large-scale temporal gene expression mapping of central nervous system development. Proc. Natl. Acad. of Sc. USA, **95(1):** 334-339.

Westhead, D.R., Parish, J.H. and Twyman, R.M. (eds) (2003). Instant Notes on Bioinformatics. BIOS Scientific Publishers Ltd., Oxford, UK.

White, K.P., Rifkin, S.A., Hurban, P. and Hogness, D.S. (1999). Microarray analysis of Drosophila development during metamorphosis. Science, **286(5447):** 2179-2184.

Wong, W.H. and Li, C. (2001a). Model-based analysis of oligonucleotide arrays: Expression index computation and outlier detection. Proc. of Natl. Acad. of Sc. USA, **98(1):** 31-36.

Wong, W.H. and Li, C. (2001b). Model-based analysis of oligonucleotide arrays: Model validation, design issues and standard error application. Genome Biology, **2(8):** research 0032.1-0032.11.

Wooley, J.C. and Lin, H.S. (2001). Catalyzing inquiry at the interface of Computing and Biology. The National Academies Press, Washington D.C., available at http://genomics.energy.gov.

Yang, Y.H., Dudoit, S., Luu, P. and Speed, T.P. (2001). Normalization for cDNA microarray data. Microarray Data Technical Report 589, SPIE BiOS 2001, San Jose, California, USA.

Yeang, C.H., Ramaswamy, S., Tamayo, P., Mukherjee, S., Rifkin, R.M., Angelo, M., Reich, M., Lander, E., Mesirov, J. and Golub, T. (2001). Molecular classification of multiple tumor types. Bioinformatics, **17:** 316S-322S.

Yeung, K.Y. and Ruzzo, W.L. (2001). Principal component analysis for clustering gene expression data. Bioinformatics, **17:** 763-774.

Yeung, K.Y., Haynor, D.R. and Ruzzo, W.L. (2001b). Validating clustering for gene expression data. Bioinformatics, **17(4):** 309-318.

Zhang, M.Q. (1999). Large-scale gene expression data analysis: A new challenge to computational biologists. Genome Research, **9:** 681-688.

Websites

Chang, C.C. and Lin, C.J. (2004). LibSVM: A library for support vector machine. Available at http://www.csie.ntu.edu.tw/~cjlin/libsvm.

Colantouni, C., Henry, G. and Pevsner, J. (2000). Standardization and Normalization of Microarray Data (SNOMAD) software. Available at http://pevsnerlab.kennedykrieger.org/snomad.htm.

de Hoon, M., Imoto, S. and Miyano, S. (2004). The C Clustering Library (Cluster 3.0) software. University of Tokyo, Institute of Medical Science, Human Genome Center, Japan, available at http://bonsai.ims.u-tokyo.ac.jp/~mdehoon/software/cluster.

Dopazo, J. (1999). Self-organizing Tree Algorithm (SOTA), DNA-array data analysis with SOM, Bioinformatics Unit at CNIO. Available at http://bioinfo.cnio.es/docus/SOTA/#Software.

Eisen Lab (1998). Cluster and TreeView software (Hierarchical clustering, k-means and tree display). Available at http://rana.lbl.gov/EisenSoftware.htm.

Johnny, R. (2002). Analysis of microarray gene expression data. Presentation slides. Available at www.kuleuven.ac.be/bio/mcb/ internet/downloads/gene_expression.pdf.

Kohonen, T., Hynninen, J., Kangas, J., Laaksonen, J. and Torkkola, K. (1996). LVQ_PAK: The learning vector quantization package. Technical Report A30, Helsinki University of Technology, Finland. Available at http://cs.hut.fi.

Leung, Y.F. (2002). My microarray journal watch. University of Hong Kong. Website available at http://ihome.cuhk.edu.hk/~b400559/1999j_mray.html.

Merelo, J.J. and Prieto, A. (1994). G-LVQ - a combination of genetic algorithms and LVQ. Available at http://geneura.ugr.es/g-lvq/g-lvq.html.

Shapiro, G. P. and Ramaswamy, S. (2002). SPSS Clementine microarray Clementine Application Template (CAT). Presentation slides, available at http://www.spssscience.com.

SilicoCyte (2004). SilicoCyte v 1.3 software. Available at http://www.silicocyte.com.

Stanford Biomedical Informatics (2004). Cleaver 1.0. Helix Bioinformatics Group, Stanford School of Medicine, Stanford University, USA. Available at http://classify.stanford.edu.

Thomas, C. (2001). CISC 873, Data Mining Notes: What is Clustering? Lecture Notes, Queen's University. Available at www.cs.queensu.ca/home/ thomas/notes/basic_association.html.

Tom Sawyer (2003). Tom Sawyer software Image Gallery. Website available at http://www.tomsawyer.com/gallery/gallery.php?printable=1.

More suggested literature and website resources

Altinok, A. (1998). Adaptive pattern classification: Kohonen SOM and LVQ1, Presentation slides.

Bentley, P.J. (2001). Digital Biology. Simon & Schuster, New York, USA.

Brazma, A. and Vilo J. (2000). Gene expression data analysis. Mini Review, FEBS 23893, FEBS Letters 480, Elsevier Science.

Butte, A. (2002). The use and analysis of microarray data, Nature Reviews, Drug Discovery, **1**: 951-960.

Chiang, D.Y., Brown, P.O. and Eisen, M.B. (2001). Visualizing associations between genome sequences and gene expression data using genome-mean expression profiles, *Bioinformatics*, **17**: 49S-55S.

Collobert, R. and Bengio, S. (2001). SVM Torch: Support vector machines for large-scale regression problems. *Journal of Machine Learning Research*, **1**: 143-160.

Cooper, M.C. and Milligan, G.W. (1988). The effect of error on determining the number of clusters. Proc. of the International Workshop on Data Analysis, Decision Support and Expert Knowledge Representation in Marketing and Related Areas of Research, 319-328.

Cornell University, Web site. Available at http://syntom.cit.cornell.edu/chips.html.

Cortes, C. and Vapnik, V. (1995). Support Vector Networks. *Machine Learning*, **20**: 1-25.

Davoli, R. (2001). Neural Networks. Dept. of Computer Science, University of Bologna, Italy, Presentation slides.

Dougherty, E.R., Barrera, J., Brun, M., Kim, S., Cesar, R.M., Chen, Y., Bittner, M. and Trent, J.M. (2002). Inference from clustering with application of gene expression microarrays, *Journal of Computational Biology*, **9(1):** 105-126.

DeRisi Lab. Department of Biochemistry & Biophysics, University of California at San Francisco. Website available at http://www.microarrays.org.

DNA Microarrays (a). Web site available at http://www.biologie.eus.fr/en/genetiqu/puces/microarraysframe.html.

DNA Microarrays (b). Web site available at http://dnamicroarrays.info.

Duggan, D.J., Bittner, M., Chen, Y., Meltzer, P. and Trent, J.M. (1999). Expression profiling using cDNA microarrays. *Nature Genetics*, **21(1 Suppl.):** 10-14.

European Bioinformatics Institute (EBI). EBI website available at http://www.ebi.ac.uk/microarray.

Ewing, R.M., Kahla, A.B., Poirot, O., Lopez, F., Audic, S. and Claverie, J.M. (1999). Large-scale statistical analyses of rice ESTs reveal correlated patterns of gene expression. *Genome Research*, **10:** 950-959.

Fuente, A. de la and Mendes, P. (2003). Integrative modeling of gene expression and cell metabolism. *Applied Bioinformatics, Open Mind Journals*, **2(2):** 79-90.

Gene Expression Data Analysis (GEDA) Tool. GEDA software. University of Pennsylvania MC Health Systems. Available at http://bioinformatics.upmc.edu/GE2/GEDA.html.

Gene Expression Pattern Analysis Suite (GEPAS) v1.0. the SOM Server. Available at http://gepas.bioinfo.cnio.es.

GenomeWeb LLC (2005). Microarray Innovators. **I.** Available at http://www.bioarraynews.com.

Gibas, C. and Jambeck, P. (2001). Developing Bioinformatics Computer Skills. O'Reilly & Associates, CA, USA.

Gutkhe, R., Schmidt-Heck, W., Hahn, D. and Pfaff, M. (2000). Gene expression data mining for functional genomics using fuzzy technology. *In:* Intelligent Applications in Biomedicine, Advances in Computational Intelligence and Learning.

Hollmen, J., Tresp, V. and Simula, O. (2000). A learning vector quantization algorithm for probabilistic models. Proc. of EUSIPCO 2000, Vol. II.

Joachims, T. (1999). Support Vector Machines (SVMlight) software. Available at http://svmlight.joachims.org.

Kim, H. (2002). Microarray analysis II: Whole-genome expression analysis. CISC889: Bioinformatics course. Presentation slides, available at www.innu.org/~super/dnac/microarray.ppt.

Lesk, A.M. (2002). Introduction to Bioinformatics. Oxford University Press, NY, USA.

Makino, S., Ito A., Endo, M. and Kido, K. (1991). A Japanese text diction recognition and a dependency grammar. *IEICE Transaction*, **E 74(7):** 1773-1782.

Milligan, G.W. and Cooper, M.C. (1985). An examination of procedures for determining the number of clusters in a data set. *Psychometrika*, **50:** 159-179.

Mitchell, T.M. (1997). Machine Learning. McGraw Hill International Edition, New Delhi, India.

Molmine. J-Express Pro 2.6 software (Hierarchical clustering, self-organizing maps, and principal components analysis), University of Bergen. Available at http://www.molmine.com.

National Cancer Institute (2002). Gene Expression Data Portal (GEDP). National Institutes of Health, USA. Available at http://gedp.nci.nih/gov/dc/servlet/manager.

National Centre for Biotechnology Information (2002). Gene Expression Omnibus (GEO). National Institutes of Health, USA Gene expression datasets available at ftp://ftp.ncbi.nih.gov/pub/geo/data/gds/soft.

Prasad, T.V. and Ahson, S.I. (2003). Labeling gene expression data using vector quantization. Proc. of 3rd RECOMB Satellite Conference, Stanford University, USA.

Prasad, T.V. and Ahson, S.I. (2004). Analysis of microarray gene expression data, Proc. of 2nd Intl. Conference on Artificial Intelligence Applications in Engineering and Information Technology (ICAIET), Universiti Malaysia Sabah, Malaysia.

Sasik, R., Hwa, T., Iranfar, N. and Loomis, W.F. (2001). Percolation clustering: A novel approach to the clustering of gene expression patterns in Dictyostelium development. Pacific Symposium of Biocomputing, 335-347.

Shamir, R. and Sharan, R. (2002). Algorithmic approaches to clustering gene expression data. *In:* Current Topics in Computational Molecular Biology (eds. Jiang et al.), 269-300, MIT Press.

Sturn, A., Quackenbush, J. and Trajanoski, Z. (2002). Genesis: cluster analysis of microarray data. *Bioinformatics*, **18(1):** 207-208.

Pat Brown Lab. Stanford University. Website available at http://cmgm.stanford.edu/pbrown/.

Reich, M., Ohm, K., Tamayo, P., Angelo, M. and Mesirov, J.P. (2004). GeneCluster 2.0: An advanced toolset for bioarray analysis. *Bioinformatics*. Earlier version available from Lander and Golub (1999). Whitehead Institute, MIT, available at http://www.broad.mit.edu/cancer/software/genecluster2/gc2.html.

Shi, L. (2002), Gene Chips Web site. Available at http://www.gene-chips.com.

von Heydebreck, A., Huber, W., Poustka, A. and Vingron, M. (2001), Identifying splits with clear separation: A new class discovery method for gene expression data. *Bioinformatics*, **17:** 107S-114S.

Wang, H., Yan, X. and Zhang, X. (2002). Analysis of gene expression profiles of hereditary breast cancer using different feature selection and classification methods. Available at http://www.columbia.edu/~xy56/project.htm.

Yang, Y.H., Buckley, M., Dudoit, S. and Speed, T. (2000). Comparison of methods for image analysis on cDNA microarray data. Berkeley Statistics Department, University of Berkeley, USA, Technical Report 584.

Zhu, H. and Snyder, M. (2001). Protein arrays and microarrays. Current Opinion in Chemical Biology, **5:** 40-45.

APPENDIX

Comparison of clustering techniques

Feature	Clustering method				
	SOM	HC	PCA	LVQ	k-means
Basic information					
Topology	Rectangular	Hierarchical tree	Rectangular	Rectangular	None
Technique	ANN	Correlation	Statistical	ANN	Statistical
Growing	Size fixed from the beginning	Aggregative (from bottom to top)	Size fixed from the beginning	Size fixed from the beginning	Fixed
Application	Clustering, visualization	Clustering	Pre-processing, clustering	Classification	Clustering
Classification	Good	Poor	Average	Excellent	Good
Prediction	Average	Very poor	Poor	Very poor	Average
Data constraints					
Ability to cope with noisy data such as gene expression data, stock market data	Good	Poor	Average	Good	Average
Handle large number of inputs effectively	Good	Poor	Excellent	Excellent	Poor
Ability to remember past and present statistics of data	Average	Very poor	Poor	Very poor	Average
Network characteristics					
Ability to detect data outside the domain of training data	Good	Poor	Excellent	Good	Good

(Contd.)

Memory requirements	Good	Very high	High	Good	Average
Learning speed	Average	None	None	Average	Average
Clustering parameters					
Number of clusters, can be less than specified, in case of null clusters	Fixed	Variable	Not defined	Fixed	Fixed
Definition of cluster	✓	✗	✓	✓	✓
Proportional clustering	✗	✓	✗	✗	✓
Possibility to obtain clusters at different hierarchical levels (can be done by removal of other clusters and analysis of remaining data)	✓	✓	✓	✓	✓
Computational capability					
Provide average values of profiles in the cluster	✓	✓	✓	✓	✓
Robust against noise	✓	✗	✗	✓	✓
Number of parameters	Too many	Very few	None	Too many	Many
Runtime	Linear	Quadratic	Quadratic	Linear	Linear
Range of accuracy	> 85 %	40-80 %	40-80 %	90-97 %	85-94 %
Visualization techniques					
Histogram	✗	✗	✓	✗	✗
Checks view	✓	✓	✓	✓	✓
Microarray view	✓	✓	✓	✓	✓

(Contd.)

(*Appendix Contd.*)

Feature	Clustering method				
	SOM	*HC*	*PCA*	*LVQ*	*k-means*
Whole sample	✓	✓	✓	✓	✓
Proximity map	✓	✓	✓	✓	✓
Temporal (incl. zoomed cluster view)	✓	✓	✓	✓	✓
Textual	✓	✓	✓	✓	✓
PC view	✗	✗	✓	✗	✗
Eigen graph	✗	✗	✓	✗	✗
Tree view	✗	✓	✗	✗	✗
Scatter plot	✗	✗	✓	✗	✗
Decision (or search) space	✗	✗	✗	✓	✗
Gene Ontology	✓	✓	✓	✓	✓
Well known variants	SOTA, HG-SOT, GCS, TS-SOM, Fuzzy Kohonen maps, par-SOM, LabelSOM	Single, average, complete and centroid linkage	Kernel PCA	RLVQ, GRLVQ, DSLVQ, DLVQ, etc. plus LVQ1, LVQ2 and LVQ3	*k*-means, median, *k*-medoids, *k*-modes

9 Data Mining for Bioinformatics— Systems Biology

T.V. Prasad and S.I. Ahson

INTRODUCTION

Biology, like many other sciences, changes when technology brings in new tools that extend the scope of inquiry. The invention of the optical microscope in late 1600 brought an entirely new vista to biology when cellular structures could be more clearly seen by scientists. Much more modern and recent electron microscope developed in the 60's enhanced the visualization of cells considerably. The application of computing to biological problems has created yet another new opportunity for the biologists of the 21st century. As computers continue to change the society at large, there is no doubt that several years of development in databases, software for data analysis, computational algorithms, computer generated visualization, use of computers to determine structures of complex bio-molecules, computational simulation of ecosystem, analysis of evolutionary pathways and many more computational methods have brought several new dimensions to biology. The technological revolution, from an ordinary computer to high-performance/ grid computing, the processes have further automated, and led to flooding of data which cannot be handled properly, due to lack of proper standards at proper time. Billions of records are being pushed by the researchers and scientists into the data repositories across the world.

However, there is a positive thing associated with this. Once the characteristics or various genes, proteins or metabolomes are known, the next step is to draw the relationship between them so that the entire picture of the "system" function can be known. Now the study of higher levels of activities within organisms has become a reality. With the data related to molecular composition of virus and other disease-causing agencies, scientists can understand the pattern of diseases like HIV, cancer, etc. It has become easy to understand the nutritional and metabolic activities inside the body, now called nutri-genomics and metabolomics respectively. It has also opened up ways to understand the impact and decomposition due to toxic compounds, now called toxico-genomics. The effect of pharmaceuticals/drugs on tissues and entire human body down at the level of genes and proteins can also be studied, now called pharmaco-genomics.

All these fields of studies lead to one special challenging area called the systems biology. It will be too early to describe it in great detail but it is essential to know various issues that are known. The growth is so high that a world map of websites of biological data would have to be in place.

The field of systems biology has many similarities to other engineering disciplines. The foundations of signaling and transmission in human body or any other organism has analogy with the electrical engineering and so also the feedback control mechanism has with the electronics and communication engineering. The modeling and simulation is a very close associate of the cartoons and animation sector, which needs a fairly high level of visualization than any other field.

Lack of proper analysis tools, methods of acquisition/accessing and representations, technical standards, and visualization techniques are the prime-most issues in handling biological data at large. Initially, it was not thought that the situation could turn out to be so cumbersome, but as the data is growing hour-after-hour, it is becoming even more difficult to integrate databases, models, or even compare the data.

Presently, there are almost no methods for integration of data, analysis and visualization tools, standardized models for computation or comparison, etc. for handling genomic data. Leading researchers and research organizations have laid roadmaps, blueprints, guidelines and so on for taking up new work but almost all of them prove to be worthless in wake of absence of qualified and trained researchers, directions and funding. American, European and Asian countries are pouring in huge funds to tackle the problems, but seem to have achieved little success.

For example, the ASN.1 (Abstract Syntax Notation) is being used only by few well-known organizations, whereas other agencies—well established and amateur—developed their own standards and generated highly voluminous yet very useful databases. This led to incompatibility and inconsistency with other databases across the world.

Scientists and researchers have numerous questions to be answered in the wake of these technological developments in information sciences and discoveries in bio-sciences. How the entire set of genes coordinate together in the development of human body? How do the genetic networks vary amidst different persons? What makes us separate from other living beings? What is the "blue print" of each person? Will the genetic information be shared publicly in future? Will it be used to clone human beings? What is the level of security? How will the data be used to find cures for serious diseases? The questions are so many and the data being gathered at the dawn of the twenty first century is so enormous that it may likely take almost the entire century to reach to a satisfactory level of understanding.

DATA MINING IN SYSTEMS BIOLOGY

Understanding and harnessing the power of 'biological systems' using systems approaches will open doors to the development of far-reaching applications of computing in the 21st century. It will be possible to use blood or tissue sample to predict an individual's susceptibility to large number of afflictions. It will be possible to propose chemotherapy regime using modeling techniques for assessing the biological response to a proposed regime. Many other research areas will utilize the systems approach and aided by the advanced computation techniques and cyber infrastructure.

Although biology changed drastically in the later years of the 20th century with the impact of new developments in molecular biology, the three strands of inquiry still continue: (a) empirical observation, (b) evolution, and (c) taxonomy of life. Biology is an empirical and descriptive science. The culture of bioscience is based on visualization—identifying new species, describing physical appearances, environment, life cycle and diversity of life. Bioscience is also concerned with taxonomy and classification. The emerging discipline of semantic web and ontology directly apply to biology. One of the main aim of biological inquiry in the 21st century is the understanding of the underlying mechanisms of life. The modern trend is systems-level understanding of structure, dynamics and control mechanisms at the organism level.

Research in bioscience has undergone a paradigm shift with the focus on cellular studies. This shift has been driven by the increasing availability of high-throughput (HT) data. The human genome project was the catalyst for developing HT technologies. The earliest form of HT data were genome and proteome sequences. Many new forms of biological data including gene expression, protein expression, metabolmics, mass spectrometry, protein structure, radiological data, are now available. The increasing quantity and quality of data has led to data curation, annotation, and integration. In order to understand and derive meaning from data various data mining techniques have been developed.

Much beyond the complex issues of data lays the roadmap towards processing biological data of varied forms. This would lead ultimately to the processing of complete human system or any other meaningful biological system. Such a system could be whole genome, genome to genome comparison or understanding the behaviour of various organism growths.

Broadly speaking, all these systems will comprise study of systems biology. It may be too early to talk about systems biology at a time the constituent fields are in infancy stage. In today's world, the in silico drug design is considered a hot discipline, whereas not many tools are available. Protein folding, docking of drug compounds into the targets, identifying the markers, etc. are the buzzword of the day.

The ever-growing types and size of databases require tools for exploring certain hidden patterns in data. The traditional statistics-based techniques, based on a priori hypothesis, can be used to discover new, interesting and previously unknown patterns in large amounts of data. The common data mining tools are based on regression, clustering, classification, association rules, decision trees and other machine learning algorithms. The fields of data mining, AI, systems biology, ontology, and database theory are becoming more interrelated while striving to provide answers to biological questions through computational means.

Data mining algorithms are applied when databases are too massive and are of various formats across different hierarchies of biological information i.e., DNA, RNA, proteins, macromolecular complexes, signaling networks, cells, organs, organisms, and species. There is a necessity to explore as to how the techniques of data mining are applied to biological databases in order to extract patterns which are crucial in making decisions in healthcare. The aim of this course is to discover, train, encourage, and support the new kinds of biologists needed for the 21st century to provide learning opportunities at the interfaces among biology, mathematics, statistics and computer science. Illustrative research areas at the interface of computer science, systems science, AI and biology are:

- Structure determination of biological molecules and complexes
- Simulation of protein folding
- Whole genome sequence assembly
- Whole genome modeling and annotation
- Full genome-to-genome comparison
- Rapid assessment of polymorphic genetic variations
- Relating gene sequences to protein structures
- Relating protein structure to function
- In silico drug design
- Modeling of cellular process
- Simulation of genetic-regulatory networks
- Modeling of physiological systems in health and disease
- Modeling behaviour of schools, swarms and their emergent behaviour
- Synthetic biology
- Artificial life
- Implantable neural prosthetic

We have already understood about the data types and issues associated with them in the preceding article titled "Data Mining for Bioinformatics – Microarray Data" as an introduction. Now, let us see the complete picture of our understanding about biological data forms and their issues (Jagadish and Olken, 2003). The biological data can be in various forms as given in Table 1.

Table 1: Different forms of biological data

Biological data type	Example
Sequences	DNA sequences of various organisms like fruit fly, mice, C. elegans
Graphs	Metabolic pathways, genetic networks, signaling pathways
High-dimensional data	Microarray gene expression data
Geometric information	Protein structure, docking of drug candidate in a binding site
Scalar and vector	Molecular models related to the cells, hydrophobicity, electricity/signal transmission/communication model in a cell
Images	Radio-logical images, X-rays, animations, etc.
Prose	Lot of biological material that contain important knowledge but running in the form of paragraphs and essays

Within a system with four parts, there are 11 possible modes of interaction. Among a class of 20 students there can be 190 possible interactions, counting just the pairwise interactions. And among the approximately 25,000 genes that comprise each human being, there are more than 336 million possible pairwise interactions. Since genes interact in more than pairs, the total number of possible interactions is staggering (ISB, 2007). Further, with thousands of researchers working on millions of organisms generating numerous forms of data using many experimental techniques, the complexities (for quality) of biological data have been laid out in the following manner:

- High volume
- Accuracy and consistency
- Organization
- Sharing
- Integration
- Curation

The computational scientists have to adopt the systems approach to solving such challenging problems, and additionally, the following parameters will also have to be addressed keeping in view the complexity:

- Interfacing/formats/standards
- Retrieval
- Validation
- Visualization
- Modeling/simulation
- Maintenance
- Documentation
- Integration and portability

Methods, models, algorithms, databases, visualization and other parameters are required for handling data right from the stage of preparation, acquisition, finalization, conversion, processing, analysis, visualization and then further use on other processes/activities. The process may start with the preparation of samples for NMR, MS, microarray, X-ray crystallography, 2D-PAGE, or any other method, all of which generate data of different formats across different platforms.

Kitano in 2003 described that a large portion of the current day's knowledge of genetic regulatory networks or systems biology is represented in the form of cartoon like networks (SBI Japan, 2003). Such knowledge cannot be represented in the form of general text or in the form of general database formats. For graphical interfaces, the following criteria could be helpful:

- Expressing the relationships between different entities such as relationship between genes and proteins
- Clarity in expression – there has to be visual and grammatical/semantic clarity in expressing the entities
- Expandability – the graphical representations should be expandable upon addition of new knowledge as and when required
- Mathematical modeling – it should be possible to derive mathematical models corresponding to the available knowledge, at any given time
- Software support – due to the bulkiness of knowledge, it should be possible to handle the operations through proper computing support

Interestingly, no graphical model today satisfy all these criteria; however, some of the available systems satisfy some of them. This is a noteworthy item for all scientists, academicians and researchers for undertaking projects in this area. A description of various types of models with suitable examples is given in Table 2.

Table 2: Various types of models used in systems biology

Models	Examples
Probabilistic graphical models	Hidden Markov models, Bayesian network models
Logical models	Boolean models, fuzzy logical models, rule-based models
Statistical models	Linear models, neural network models, Bayesian network models
Computational models	Simulation models
Multi-scale models	Molecular interaction models
Power law models	Generic power model, general theory of power law scaling
Hybrid models	HMM with partial differential equations, simulation with differential equations
Qualitative models	
Quantitative models	

MODELING AND SIMULATION

ISB (2007) envisages that the most interesting aspect about systems biology is the area of predictive, preventive and personalized (PPP) medicine, whose goals include:

- Identification and classification of diseases and patient populations so that the diagnosis becomes more specific and treatment more effective. Till date, the diseases were seen as cellular imbalances, which should otherwise be classified in terms of the cell-type specific biological networks that have been perturbed by the defective genes and/or environmental stimuli.
- More flexible drug designing, preferably in silico, based on computer simulations of docking of drug compounds in a complex protein or gene regulatory network, with an aim to enhance the efficiency and reducing the side effects.
- To make best use of genetic information to evaluate a person's health history in real time. The blood can become a window into screening of health and disease.
- Restoration of a disease-perturbed network to its normal state by genetic or pharmacological intervention, with compounds that are specially developed for the patient and to repair the diseased networks. Also, the idea is to identify changes in cellular networks at the earliest possible stage and, to find a solution to prevent or limit the effects of the disease.

ISB also envisages that PPP has the potential to transform traditional medicine to considerably decrease the effect and mortality of serious diseases like cancer, Huntington's, Parkinson's and diabetes. Other applications will definitely increase the personal and communal well-being as the detrimental effect of abnormalities associated with mental illness will be eased.

Some of the oldest references related to the modeling and simulation of the genetic regulatory networks are McAdams and Shapiro (1995), Somogyi and Sniegoski (1996), Goldsby et al. (1999) and Forst (1999).

Aravind (2000) used association models for genome analysis, whereas D'haeseler et al. (2000) used genetic network inference to arrive at reverse engineering from co-expression clustering. Other noted works are Diehn and Relman (2001) and Eisenberg (2000). Forst (2001) is one of the first tutorials in systems biology. Wiechert (2002) termed a new phrase "metabolic engineering" in his work and described various tools for modeling and simulation used in it. The concepts of electrical engineering, control systems theory, physics, graph and network theory, etc. are being applied to the regulatory networks as well as other forms of pathways. Wolkenhauer (2002) gave a very lucid explanation of comparing mathematical modeling concepts with those applicable to genetic networks. Ullah and Wolkenhauer (2007) is a very useful literature that describes in detail the family tree of the Markov models that can be applied in systems biology.

The study of systems biology though deeply rooted in graph theory and on various simulation and modeling techniques, find implementation in a number of areas. Wooley and Lin (2001) suggested the roadmap for applications in the future in the following areas:

1. Molecular and structural biology
   (a) Prediction of complex protein structures
   (b) Protein folding
   (c) Identification/recognition of binding sites/markers (functional and structural)
   (d) Clustering/classifying functional class of proteins
   (e) Molecular docking, in silico
2. Cell biology and physiology
   (a) Cellular modeling and simulation, pathway reconstruction
   (b) Cell cycle regulation
   (c) Inhomogeneities in cellular development and signaling
3. Genetic regulation
   (a) Regulation of transcription activity as process control computing
   (b) GRN as finite state automata
   (c) GRN as electrical circuits
   (d) Combinatorial synthesis of GRN
   (e) Combining experimental data with biological network information
4. Organ physiology
   (a) Multi-scale physiological modeling
   (b) Hematology for leukemia
   (c) Immunology
   (d) Modeling of the cardiac system
5. Neuroscience
   (a) Landscape of computational neuroscience
   (b) Large scale neural modeling
   (c) Muscular control
   (d) Synaptic transmission
   (e) Neuropsychiatry
   (f) Recovery from injury
6. Virology
   (a) HIV/AIDS
   (b) HPV
7. Epidemiology
   STD
8. Evolution and ecology
   (a) Study of commonalities
   (b) Characteristics of evolution
   (c) Characteristics of ecology

Before actually working on complex biological systems such as human and livestock, simpler organisms, especially the single cell organisms such as yeast and bacteria could be modeled. With fewer genes and simpler genetic regulatory network, it will undoubtedly become far easier to work with, analyze and predict.

Just as it happens in the agricultural and meteorological forecasting, the models are required to be tested stringently over and again for many biological cycles before stable and more accurate models are arrived at. This rigorous exercise may take many such cycles over long periods, even with the use of best tools and techniques. The outcomes from experimental perturbations during such exercises can also yield very interesting results.

VISUALIZATION IN SYSTEMS BIOLOGY

In addition to the basic visualization forms described in the earlier article (Data Mining for Bioinformatics – Microarray Data), a variety of other visualizations for microarray gene expression data are available, at websites of GEDA and SilicoCyte, which can be easily converted and/or presented for biological data representation:

- 2D and 3D score plots
- Profile Plots
- Scatter plots
  - 3D scatter plots
  - PCA visualization - Result on 3D scatter plot
  - Isomap visualization
  - Multi-dimensional scaling
- Venn Analysis Diagrams for visualizing similar elements in microarrays, same genes in gene groups or same genes in microarray designs
- SOM visualization of clustering result on
  - 2D Trellis plot
  - U-matrix
  - 1D SOM

There are a number of visualization techniques to represent interaction between the genes in a genetic regulatory network. The mapping could be interaction between protein-protein, protein-DNA, protein-RNA, protein-ligand bindings, DNA-binding ligand, post-translational modifications of proteins, metabolites, etc. One of such excellent work of visualization is given by Sawyer (2003), where a number of methods of presenting different forms of outputs are given as examples, and most of which can be used to indicate regulatory networks, pathways or interaction networks very comfortably.

In other words, these network types have been more or less standardized. Broadly speaking, though all these visualizations are based on the graph theory, there are three forms of visualization techniques for representing the network structure of the biological outputs, viz., the hierarchical tree structure, the forest structure (consisting of a number of tree/graph like structures), the circuit diagram structure (very useful for representing haphazard and complex relationships between entities) and a hybrid structure. For the sake of clear conceptualization, some of these visualizations have been reproduced in Fig. 1. Please refer to the website for clear details.

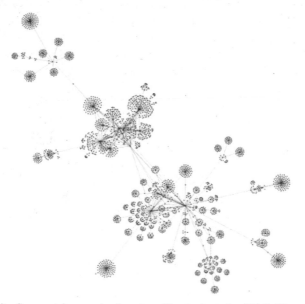

Fig. 1 (a): Symmetric graph drawing illustrating the BBC News Agency website map. Similar visualization would be useful for representing linkages of genes/proteins/metabolomes interactions.

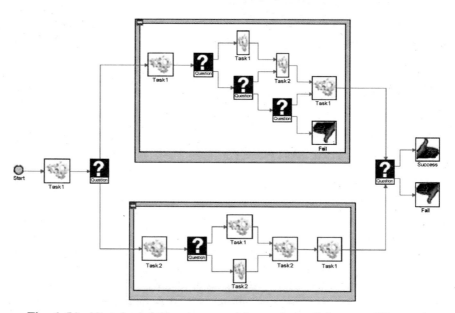

Fig. 1 (b): Nested graph drawing organizing and visualizing a workflow task process. When it is desired to indicate the connectivity of different networks together to give a complete view of the regulatory network or pathways, this visualization can be applied.

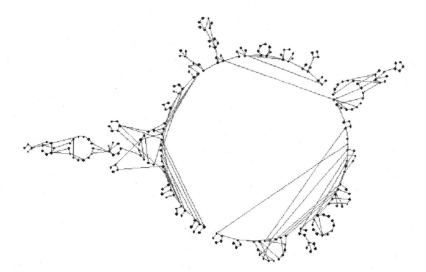

Fig. 1 (c): A circular graph drawing revealing networks of cluster trees radiating from a central hub cluster. There are requirements to plot the entire cycle of conversion from one phase to another in the genetic regulatory network, which also helps in indicating the circular relationships of sub-components or sub-networks.

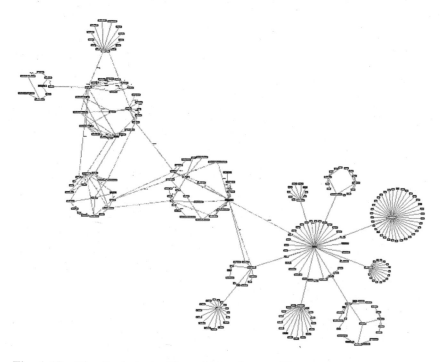

Fig. 1 (d): Circular layout with multiple clusters. Clusters are a very common feature in genomics and proteomics. Indicating individual clusters as well as their relationship with others can be as effective as this visualization.

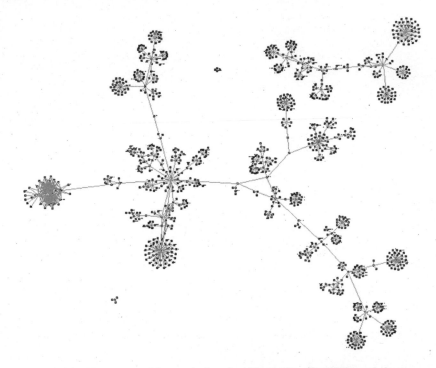

Fig. 1 (e): Vast symmetric graph drawing illustrating the CNN website map. This visualization, called a forest (group of trees), can be compared to a larger representation of connectivity between various clusters and individual entities (genes, proteins, metabolomes or others, as the case may be) in systems biology.

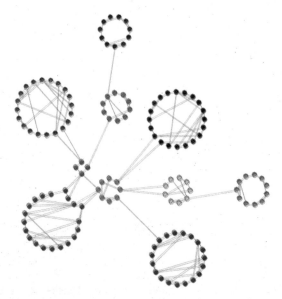

Fig. 1 (f): Circular graph drawing with multicolour clusters. Using different colour codes for genes or proteins of similar functionality can be a very good idea.

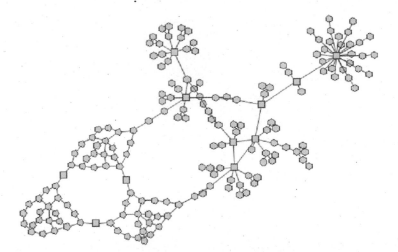

Fig. 1 (g): Symmetric graph drawing representing objects and their complex relationships clearly. The relationships between genes, proteins and transcriptomes need not also be as simple as generally visualized. They may involve a number of stages and regulations as shown in this visualization.

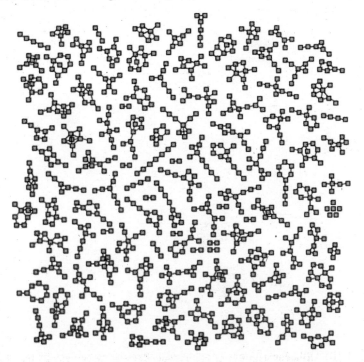

Fig. 1 (h): This graph drawing demonstrates the efficiency of the packing algorithm, which tightly places groups of connected components. It is important to realize that in practice the genes and protein regulation is a set of disjoint trees, or forest. For successful systems biology visualization, the software should consider this aspect.

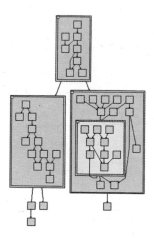

Fig. 1 (i): Orthogonal graph drawing organizing classes and their relationships. In case it becomes difficult to indicate all the intricacies of the regulatory network or pathways, the relationships and transcriptome could be represented as a set of inter-connected graphs as shown here.

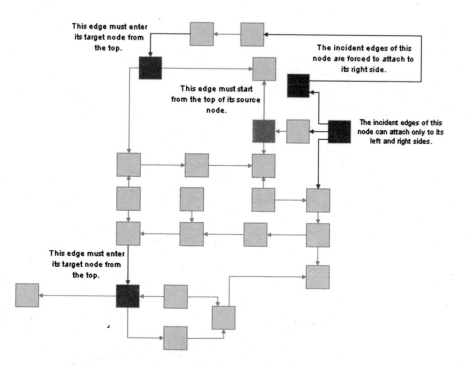

Fig. 1 (j): Orthogonal graph drawing indicates the placement of the source and target nodes. Labeling functionality accommodates the space needed to label the drawing.

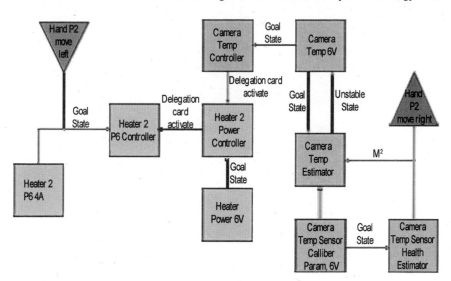

Fig. 1 (k): An orthogonal graph drawing taken from the domain of robotics design shows how hyper edges (edges between more than two nodes) can be rendered and laid out. This is one of the simplest forms of visualization which can be easily implemented for a smaller genetic regulatory network or protein interaction map. The effect of catalyzing inputs over the transcription can be indicated using this method, as given by the initiator/terminal node (triangles).

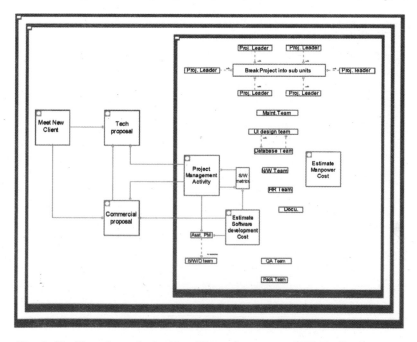

Fig. 1 (l): Nested graph drawing illustrating a nested UML diagram. For zooming in and out any complex regulatory network, the nested graph as shown above could be very useful.

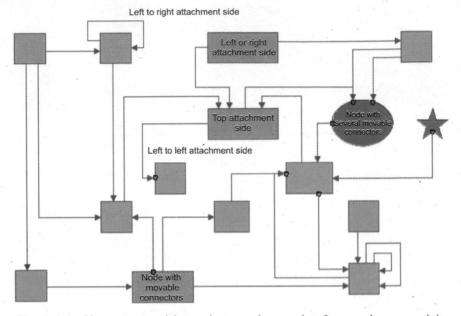

Fig. 1 (m): You can control how edges attach to nodes. In case the connectivity are to be shown on different pages, or stages are to be shown, this visualization can help in organizing the output in the most effective manner.

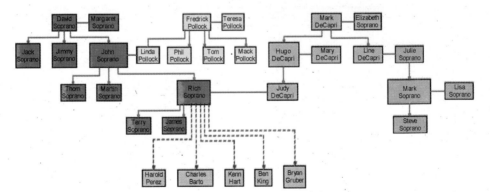

Fig. 1 (n): Hierarchical graph drawing depicting the Sopranos family tree. It uses coloured nodes to differentiate separate lineages. Orthogonal edge routing creates a clear organizational view and horizontal edge routings illustrate couples. For a very simpler regulatory network representation, this visualization can be employed.

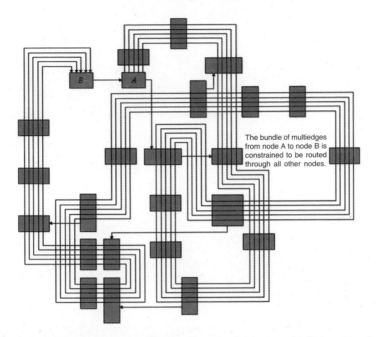

The bundle of multiedges from node A to node B is constrained to be routed through all other nodes.

Fig. 1 (o): The bundle of multiedges from node A to node B is constrained to be routed through all other nodes. The transcriptions may involve various regulations initiated simultaneously. In the above visualization, bundle of connections represent the flow of information between two stages.

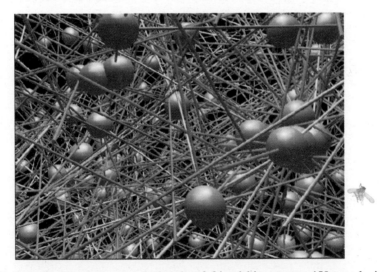

Fig. 1 (p): Social graph records bonds of friendship among 450 people in Canberra, Australia. The network of social contacts was documented by Alden S. Klovdahl of Australian National University. Image was created with View_Net, written by Klovdahl and R.H. Payne; refer Hayes (2000). The social contacts network is in all ways similar to the regulatory networks and finds straight application.

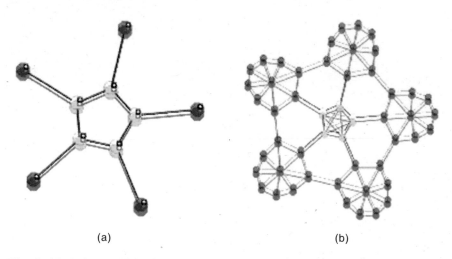

(a)   (b)

Fig. 2 (a) *Before* – This simply structured symmetric graph drawing is hiding a more complex structure under each of its nodes. (b) *After* – Entire symmetric graph drawing after the hidden nodes and edges are revealed.

SYSTEMS BIOLOGY – TOWARDS MOLECULAR PATHWAYS

Reconstructing molecular pathways from expression data is a difficult task. One approach is to simulate pathways using a variety of mathematical models and then choose the model that fits the data. Reverse engineering is a less demanding approach in which models are built on the basis of the observed behaviour of molecular pathways. It is possible to represent the gene expression data by using simultaneous differential equations or representing the expression as Boolean networks; however, each method suffer from disadvantages. A hybrid model such as the finite linear state model could bring good results (Westhead et al., 2003).

Gene expression and disease link: Gene expression data analysis could give important clues for reasons behind cause of various "third world" diseases, e.g., malaria, typhoid, cholera, HIV/AIDS, cancer, dengue, etc., especially about the behaviour and growth of disease causing germs.

Smaller datasets are preferred for processing: Large samples of over 200 genes are not helpful since if all of them are correlated with a particular class, it is unlikely that they all represent different biological mechanisms and hence are unlikely to add information not already provided by others. Though excellent research work has been carried out in last three decades, there are no general approach for identifying new cancer classes (class discovery) or for assigning tumours to known classes (class prediction) (Jagota, 2001).

    Most of the data in bio-sciences pertain to textual form or numeric form rather than any symbolic or image form. That is why, data mining of biological nature, especially the gene expression, sequence data and other forms of non-symbolic data is considered to be efficient and easier than any other forms. Higher level of data types could be used while describing the protein structure data or genetic regulatory networks. However, similar experiments and exercises can be conducted to study the effect of various chemicals, germs, micro-organisms, drugs, etc. on the human body or on other creatures during clinical trials. With the massive reduction in the cost of microarray experiments, it is expected that the biochemistry and pathological laboratories would be flooded with extremely massive data and most of the chemicals and experimental procedures would straightaway be replaced by computers, microarrays, data mining algorithms under the title in silico.

New areas such as metabolomics, toxico-genomics and pharmacogenomics have sprung up in which gene expression data continues to be the back-bone framework. In a recent review by de la Fuente and Mendes (2003), works related to application of various methods of modeling both gene expression and cell metabolism together have been described in great detail. The three major entities transcript, protein and metabolite form the larger framework whereas the enzyme kinetics are responsible for the interactions down below at the chemical network level. When these chemical networks are analyzed either at individual (micro) level or of the entire process from transcript to metabolite, features of various constituent elements could be deciphered.

Table 3: Comparison between various concepts in real life and those in systems biology

Entity	Similarity in systems biology
Semantic web	Neurological network
	Infrastructure for World Map of websites
Petri-nets	GRN as circuits
Trees	Phylogenetic trees, ancestral trees, hierarchical clustering
Finite state automata and/or Boolean networks	Genetic expression pattern in GRN

The most pressing challenges of systems biology fall roughly into the following four categories, ISB (2007):

- *Experimental* — pertains to strategies for designing experiments and collecting reliable data.
  Technological — pertains to the development of new instrumentation for making rapid, highly parallel, inexpensive and accurate measurements of informational molecules and their sequence, structure, modifications or processing, localization and interactions with other components large and small.
- *Computational* — pertains to the development and refinement of network theory and effective engineering of simulation tools, so that descriptive

networks can be replaced by more accurate dynamic models of the system's molecular interactions.
- *Sociological* — pertains to effective communication across disciplines, the dynamics of research teams, difficulties obtaining funds, and the like.

Of these computer scientists would like to address the computational challenges of systems biology. The ISB has formulated 11 computational or mathematical challenges in contemporary biology:

1. How to fully decipher the (digital) information content of the genome?
2. How to do all-versus-all comparisons of thousands of genomes?
3. How to extract protein and gene regulatory networks from the challenges 1 and 2 as indicated above?
4. How to integrate multiple high-throughput data types dependably?
5. How to visualize and explore large-scale, multi-dimensional data?
6. How to convert static network maps into dynamic mathematical models?
7. How to predict protein function *ab initio*?
8. How to identify signatures for cellular states (e.g. healthy vs. diseased)?
9. How to build hierarchical models across multiple scales of time and space?
10. How to reduce complex multi-dimensional models to underlying principles?
11. How to do text searching to bring the literature and experimental data together?

A large number of areas of research have sprung up due to the need to understand the functions of various entities inside the body of an organism. The areas of study not only got added to the biosciences but to the information sciences as well. Some of them are as follows:

Systems biology as source for computing

- Swarm intelligence
- Robotics
- Self healing systems
- Immunology and computer security
- Amorphous computing

Systems biology as implementer for computing

- Evolutionary computing
- Robotics – energy management
- Neuroscience and computing
  - ◆ NN
  - ◆ NN inspired sensors
- Ant algorithms

Systems biology as physical substrate for computing

- Biomolecular computing
- Synthetic biology
  - Artificial life
  - Cellular logic gates
- Nano-fabrication and DNA self-assembly

SYSTEMS BIOLOGY RELATED SOFTWARE

Systems Biology Markup Language (SBML) is a language developed for representing information and exchanging information related to biological systems. It also helps in modeling various biological systems and is based on XML implementation. It was developed by the Caltech ERATO Kiranto Systems Biology Project Group, Caltech.

Details of some software based on SBML are given below; refer Funahashi (2003) for more details:

- SBML compliant, SBW-enabled application
  - Visual editors – CellDesigner, JDesigner
  - Simulators – Jarnac, Gepasi, Gibson simulator, Trelis
  - Others – MathSBML, MATLAB translator, libSBML
- Converters
  - KEGG to SBML converter
  - DOQCS to SBML converter
- Jarnac is a biochemical simulation package for Windows written and maintained by Herbert M. Sauro at the Keck Graduate Institute http://www.sys-bio.org/Jarnac.htm
- Gibson Simulator is a simple stochastic simulator based on Gibson-Bruck variant of the Gillespie algorithm and is maintained by SBW development group.
- JDesigner is also a visual editor for designing biochemical network layout similar to the CellDesigner. It is written and maintained by Herbert M. Sauro at the Keck Graduate Institute and is available at http://www.sys http://www.sys-bio.org/JDesigner.htm.
- Schmidt and Jirstrand (2005) described that the Systems Biology Tool Box (SBTOOLBOX for MATLAB) offers an open and customizable environment, which can be used by scientists and researchers to explore new ideas, prototypes and new algorithms based on the requirements. It also allows building of application models for the analysis and simulation of biological systems. It provides a number of features:

  - Experiment description functionality
  - Modeling (based on equations of chemical reactions)
  - Handling of measurement data

- Import of SBML models from other software
- Simulation (deterministic and stochastic)
- Steady-state and stability analysis
- Determination of the stoichiometric matrix
- Parameter sensitivity analysis
- Network identification
- Parameter estimation
- Simple model reduction
- Bifurcation analysis
- Localization of mechanisms leading to complex behaviours
- Optimization algorithms

- Gepasi is a software package for modeling biochemical systems. It is written and maintained by Pedro Mendes at Virginia Tech. University, and is available at http://www.gepasi.org.

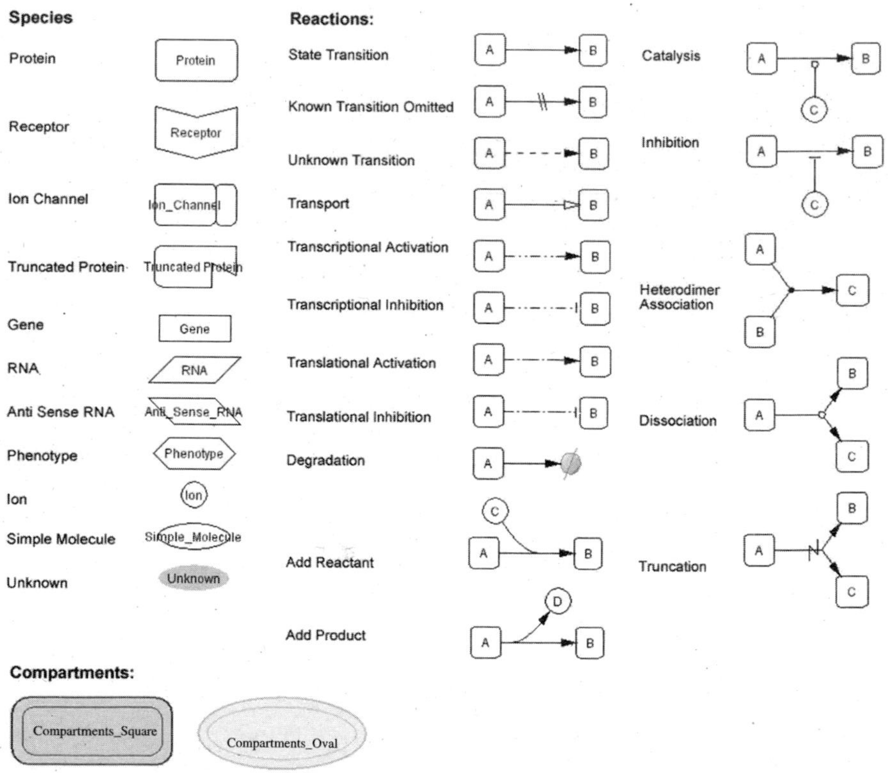

Fig. 3: Some symbols and expressions in CellDesigner software based on the proposals of Kitano for SBML. CellDesigner is freely available for download from http://www.systems-biology.org.

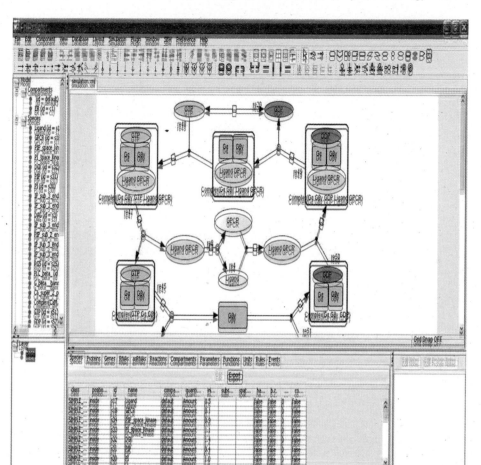

Fig. 4: The symbols and expressions can be visually laid out to form a model.

- MathSBML is also known as The Mathematica SBML Reader. It is written and Maintained by Bruce E. Shapiro at California Institute of Technology, and is available at http://www.sbml.org.
- Trelis is a graphical Monte Carlo simulation tool for modeling the time evolution of chemical reaction systems, developed by Luigi Warren at California Institute of Technology, http://sourceforge.net/projects/trelis.
- A host of translator, validation, simulation, etc. tools are available at http://sbw.kgi.edu/. Some of them are as follows:
- SBML layout viewer: displays/creates a layout for SBML models (Fig. 4)
- SBML validation: validates SBML models
- SBML simulation: simulates SBML models using various simulators
- SBML translator: translates SBML to Fortran/Java/Matlab/XPP and a number of other platforms

- University of Washington's Systems Biology Workbench 2.7.6 released during Oct 2007 is available at http://sys-bio.org/sbwWiki/doku.php?id=sysbio:sbw.
- CompuCell3D is a tissue and organ simulator based on the Cellular Potts Model, available at http://simtk.org/home/compucell3d. In collaboration with the CompuCell3D team at the University of Bloomington (Glazier et al.) a combined tool was demonstrated linking CompuCell3D with SBW. The software uses Python as a flexible scripting tool that combines the Cellular Potts Model with the portable reaction-based simulator (roadRunner) available with the Systems Biology Workbench. All modules can be run on Linux as well as on Windows or the Mac OS X.
- The Institute for Systems Biology has a range of software products for different activities/analysis under categories such as data generation software, data management software, data visualization and analysis software, etc. The details are as follows:

Data Generation Software

ISB proteomics pipeline:
http://www.proteomecenter.org/software.php
ttp://sashimi.sourceforge.net/

ISB microArray pipeline:
http://db.systemsbiology.net/software/ArrayProcess/
http://db.systemsbiology.net/software/VERAandSAM/

Data Management Software

The Systems Biology Experiment Analysis Management System (SBEAMS):
http://www.sbeams.org/

Juvenile Diabetes Research Foundation's Center for Bioinformatics at ISB
http://www.T1DBase.org/

Research Community Database for Huntington's Disease
http://HDBase.org/

Data Visualization and Analysis Software

BioSap (Blast Integrated Oligonucleotide Selection Accelerator Package):
http://biosap.sourceforge.net/

BioTapestry (for editing and visualization of genetic networks in developing embryos)
http://www.BioTapestry.org
http://magnet.systemsbiology.net/

Cytoscape (for data integration, network visualization, and analysis):
http://db.systemsbiology.net/cytoscape/plugins/
http://www.cytoscape.org/

Dizzy (simulation of deterministic and stochastic kinetics in biochemical networks) :
http://magnet.systemsbiology.net/software/Dizzy/
http://magnet.systemsbiology.net/

The GESTALT Workbench for genomic sequence analysis:
http://db.systemsbiology.net/gestalt/

Mogul: transcription factor binding site prediction multi-algorithm server
http://xerad.systemsbiology.net/Mogul/
http://magnet.systemsbiology.net/

The Multiobjective Analyzer for Genetic Marker Selection (MAGMA):
http://snp-magma.sourceforge.net/

Pointillist (tools for data integration and network model building)
http://magnet.systemsbiology.net/software/Pointillist/
http://magnet.systemsbiology.net/

RepeatMasker (screens DNA sequences for interspersed repeats):
http://www.repeatmasker.org/

It is also worth notable that the BioPAX and the BioPathways Consortiums are working in the area of application of semantic web to the biological/signaling pathway representation and re-construction. Substantial amount of success has been achieved using the PathoLogic software in which semantic web technique was applied to provide meaningful predictions in *E. coli* and *S. cerevisieae*. It has also been applied on the database of marine plankton cyanobacteria, *Prochlorococcus marinus*.

CONCLUSIONS

Overall, there is a significant requirement of standardization in the microarray gene expression data analysis scenario, which currently suffers from the following limitations due to non-standardization, and consequently, it can be concluded that efforts would be needed to formalize on which platform, algorithm, technique/model, parameter, and visualization to be used for analysis of various kinds of data:

- Microarray production and techniques not standardized
- Preprocessing techniques not yet standardized
- No standardized process or model available for analysis
- No standardized visualization techniques available
- Standardization of minimal information exchange on microarray data brought out recently under the MIAME project

REFERENCES

Articles/Papers/Presentations/Books

Aravind, L. (2000). Guilt by association: Contextual information in genome analysis, Genome Research, **10:** 1074–1077.

Baldi, P. and Brunak, S. (2003). Bioinformatics: The Machine Learning Approach, 2ed. Affiliated East-West Press Pvt. Ltd., New Delhi.

D'haeseler, P., Liang, S. and Somogyi, R. (2000). Genetic network inference: From co-expression clustering to reverse engineering. *Bioinformatics*, **16:** 707-726.

Diehn, M. and Relman, D. (2001). Comparing functional genomic datasets: Lessons from DNA microarray analyses of host-pathogen interactions. *Current Opinion on Microbiology*, **4:** 95-101.

Eisenberg, D., Marcotte, E.M., Xenarios, I. and Yeates, T.O. (2000). Protein function in the post-genomic era. *Nature*, **405:** 823-826.

Forst, C.V. and Schulten, K. (1999). Evolution of metabolism: A new method for the comparison of metabolic pathways using genomic information. *Journal of Computational Biology*, **6:** 343-360.

Forst, C.V. (2001). A Tutorial on Network Genomics. International Conference on Intelligent Systems for Molecular Biology, Copenhagen, Denmark.

Funahashi, A. (2003). Introduction to SBML and SBML compliant software. ERATO Kitano Symbiotic Systems Project, presentation slides. Available at http://sbml.org/workshops/tokyotutorial/tutorial.htm.

Fuente, A. de la and Mendes, P. (2003). Integrative modeling of gene expression and cell metabolism. *Applied Bioinformatics, Open Mind Journals*, **2(2):** 79-90.

Goldsby, R.A., Kindt, T.J. and Osborne, B.A. (1999). Kuby Immunology. W. H. Freeman & Co., 4 edition.

Han, J. and Kamber, M. (2001). Data Mining: Concepts and Techniques. Elsevier, San Francisco.

Hayes, B. (2000). Graph Theory in Practice: Part 1. *American Scientist*, **88(1)**. Available at http://www.americanscientist.org/

Jagadish, H.V. and Olken, F. (2003). Data Management for the Biosciences. Report of the NSF/NLM Workshop of Data Management for Molecular and Cell Biology, Feb 2003. Available at http:/ww.eecs.umich.edu/~jag/wdmbio/wdmb_rpt.pdf.

Jagota, Arun (2001). Microarray data analysis and visualization. Dept. of Computer Engineering, University of California, CA., USA.

McAdams, H. and Shapiro, L. (1995). Circuit simulation of genetic networks. *Science*, **269:** 650-656.

Mitchell, T.M (1997). Machine Learning. McGraw Hill International Edition, New Delhi.

Schmidt, H. and Jirstrand, M. (2005). Systems Biology Toolbox for MATLAB: A computational platform for research in Systems Biology. Bioinformatics Advance Access, available at http://www.fcc.chalmers.se/~henning/SBTOOLBOX/

Somogyi, R. and Sniegoski, C. (1996). Modeling the complexity of genetic networks. *Complexity*, **1:** 45-63.

Tavazoie, S., Hughes, J.D., Campbell, M.J., Cho, R.J. and Church, G.M. (1999). Systematic determination of genetic network architecture. *Nature Genetics*, **22:** 281-285.

Ullah, M. and Wolkenhauer, O. (2007). Family tree of Markov models in systems biology. *IET Systems Biology*, **1(4)**.

White, K.P., Rifkin, S.A., Hurban, P. and Hogness, D.S. (1999). Microarray analysis of Drosophila development during metamorphosis. *Science*, **286(5447):** 2179-2184.

Wiechert, W. (2002). Modeling and simulation: Tools for metabolic engineering. *Journal of Biotechnology*, **94:** 37-63.

Wolkenhauer, O. (2002). Mathematical modelling in the post-genome era: understanding genome expression and regulation – a system theoretic approach. *BioSystems*, **65:** 1-18.

Wooley, J.C. and Lin, H.S. (2001). Catalyzing inquiry at the interface of Computing and Biology. The National Academies Press, Washington D.C. Available at http://genomics.energy.gov.

Zhu, H. and Snyder, M. (2001). Protein arrays and microarrays. *Current Opinion in Chemical Biology*, **5:** 40-45.

Websites

Biology WorkBench. San Diego Supercomputer Centre, Bioinformatics and Computational Biology group, Deparment of Bioengineering at University of California, San Diego. Available at http://workbench.sdsc.edu.

BioPathways Consortium. BioPathways website. Available at http://www.biopathways.org.

Colantouni, C., Henry, G. and Pevsner, J. (2000). Standardization and Normalization of Microarray Data (SNOMAD) software. Available at http://pevsnerlab.kennedykrieger.org/snomad.htm.

de Hoon, M., Imoto, S. and Miyano, S. (2004). The C Clustering Library (Cluster 3.0) software. University of Tokyo, Institute of Medical Science, Human Genome Center, Japan. Available at http://bonsai.ims.u-tokyo.ac.jp/~mdehoon/software/cluster.

DeRisi Lab. Department of Biochemistry & Biophysics, University of California at San Francisco. Website available at http://www.microarrays.org.

European Bioinformatics Institute (EBI). EBI website. Available at http://www.ebi.ac.uk/biomodels.

Gene Expression Data Analysis (GEDA) Tool. GEDA software. University of Pennsylvania MC Health Systems. Available at http://bioinformatics.upmc.edu/GE2/GEDA.html.

Genomes: GTL (32007). Genomes: GTL (formerly Genomes to Life Initiative), US Department of Energy. Available at http://doegenomestolife.org.

Harvard Medical School. Department of Systems Biology, Harvard Medical School. Available at http://sysbio.med.harvard.edu/.

IBM. IBM Research website. Available at http://www.research.ibm.com/grape/.

Institute for Systems Biology (ISB) (2007). ISB website. Available at http://www.systemsbiology.org.

Joint Center for Structural Genomics. Joint Center for Structural Genomics. Available at http://www.jcsg.org.

Keck Graduate Institute. Keck Computational Systems Biology (California) website. Available at http://sbw.kgi.edu/.

Massachusetts Institute of Technology. Computational and Systems Biology Initiative (MIT) website. Available at http://csbi.mit.edu.

Molecular Sciences Institute. Molecular Sciences Institute (Berkeley). Available at http://www.molsci.org/Dispatch.

National Cancer Institute (2002). Gene Expression Data Portal (GEDP), National Institutes of Health, USA. Available at http://gedp.nci.nih/gov/dc/servlet/manager.

National Centre for Biotechnology Information (2002). Gene Expression Omnibus (GEO), National Institutes of Health, USA. Gene expression datasets available at ftp://ftp.ncbi.nih.gov/pub/geo/data/gds/soft.

Ottawa Institute of Systems Biology. Ottawa Institute of Systems Biology (Canada) website. Available at http://mededu.med.uottawa.ca/oisb/eng/.

Pacific Northwest National Laboratory. Systems Biology at PNNL. Available at http://www.sysbio.org/.

Princeton University. Lewis Siegler Institute for Integrative Genomics, Princeton University. Available at http://www.genomics.princeton.edu/.

Princeton University. Center for Systems Biology, Institute for Advanced Study, Princeton University. Available at http://www.csb.ias.edu/.

SBML. Caltech ERATO Kiranto Systems Biology Project Group. Available at http://www.cds.caltech.edu/erato.

SilicoCyte (2004). SilicoCyte v 1.3 software. Available at http://www.silicocyte.com.

Systems Biology Institute, Japan (2003). The Systems Biology Institute (Japan) website. Available at http://www.sbi.jp/indexE.html, see also http://www.systems-biology.org/.

Tom Sawyer (2003). Tom Sawyer software Image Gallery. Website available at http://www.tomsawyer.com/gallery/gallery.php?printable=1.

University of Auckland. University of Auckland, New Zealand. Available at http://www.cellml.org.

University of Stuttgart. Systems Biology Group at the University of Stuttgart (Germany) website. Available at http://www.sysbio.de/.

Virginia Tech. GEPASI software website, Virginia Bioinformatics Institute, Virginia Tech. Available at http://www.gepasi.org.

10 Environmental Cleanup Approach Using Bioinformatics in Bioremediation

M.H. Fulekar

INTRODUCTION

Environmental pollutants have become a major global concern. The modern growth of industrialization, urbanization, modern agricultural development, energy generation, have resulted in indiscriminate exploitation of natural resources for fulfilling the human desires and need, which have contributed in disturbing the ecological balance on which the quality of our environment depends. Human beings in true sense are the product of their environment. Man-environment relationship indicates that pollution and deterioration of environment has a social origin. The modern technological advancements in chemical processes have given rise to new products, new pollutants and in much abundant level which are above the self cleaning capacities of environment. One of the major issues in recent times is the threat to the human life caused due to the progressive deterioration of the environment.

Technological revolution has brought new changes in products and processes in industry. The waste generated from the development of products and processes are of concern to the environmentalist. A variety of pollutants such as xenobiotics, polycyclic aromatic hydrocarbons (PAHs), chlorinated and nitro-aromatic compounds are found depicted to be highly toxic, mutagenic and carcinogenic for living organisms (Zhang and Bennett, 2005; Samanta et al., 2002). Heavy metals are also reported persisting into the environment causing toxicity to living organisms through bioaccumulation, adsorption and biotransformation. A number of microorganisms, as a result of their diversity, versatility and adaptability in the environment, are considered to be the best candidates among all living organisms to remediate most of the environmental contaminants into the natural biogeochemical cycle.

These natural forces of biodegradation can reduce wastes and cleanup some types of environmental contaminants. Composting can accelerate natural biodegradation and convert organic wastes to valuable resources. Wastewater treatment also accelerates natural forces of biodegradation, breaking down organic matter so that it will not cause pollution problems when the water is released into the environment. Through bioremediation, microorganisms are used to clean up oil spills and other types of organic pollutants. Therefore bioremediation provides a technique for cleaning up pollution by enhancing the same biodegradation processes that occur in nature.

Bioremediation of a contaminated site typically works viz. (i) to enhance the growth of whatever pollution-eating microbes might already be living at the contaminated site and (ii) specialized microbes are added to degrade the contaminants.

Bioremediation offers many interesting possibilities from a bioinformatic point of view still slightly explored. This discipline requires the integration of huge amounts of data from various sources: chemical structure and reactivity of organic compounds sequence, structure and function of proteins (enzymes); comparative genomics, environmental microbiology and so on. The accumulation of huge amounts of data on individual genes and proteins allowed the first studies of biology from a 'systems' perspective (Alves et al., 2002; Fraser et al., 2002). The bioinformatics resources devoted to bioremediation is still scarce. Some interesting projects are being carried out to organize and store this huge amount of information relating to this subject. The University of Minnesota Biocatalysts/Biodegradation Database (UMBBD) (Rison and Thornton, 2002) is among the more prominent resource. The current set of data includes 740 chemical compounds (2167 synonyms), 820 reactions, 502 enzymes and 253 organisms (Eills et al., 2003).

The microorganisms display a remarkable range of contaminant degradable ability that can efficiently restore natural environmental conditions (Ideker et al., 2001; Jeong et al., 2001). Advances in the molecular biology technologies are making a global gene expression profile possible; genome-wide analysis of DNA (genomics), RNA expression (transcriptomics) and protein expression (proteomics) as well as exploring complexes of protein aggregation such as protein-protein interaction (interactomics) create the opportunity to systematically study the physiological expressions of such organisms. Attempts are made to interpret some of the areas of genomics and proteomics which have been employed for bioremediation studies. The studies on bioinformatic data obtained viz. various sources are interpreted by combining these techniques and applying them towards future studies of active bioremediation to develop environmental cleanup technologies.

Background

The revolution in computer technology has made it possible to model grand challenge problems such as large scale sequencing of genomes and management of large integrated databases. This availability of data has led to an explosion of genome and proteome analysis leading to many new discoveries and tools. The availability of genomics and proteomics data and improved bioinformatics and biochemical tools has raised the expectation of the humanity to be able to control the genetics by manipulating the existing microbes. In addition, bioinformatic analysis has enhanced our understanding about the genome structure and the microorganisms restructuring process. Bioinformatic analyses have facilitated and quickened the analysis of systematic level behaviour of cellular processes, and understanding the cellular processes in order to treat and control microbial cells as factories for the last decade. Bioinformatic techniques have been developed to identify and analyze various components of cells such as gene and protein function, interactions, and metabolic and regulatory pathways. The next decade will belong to the understanding of cellular mechanism and manipulation using the integration of bioinformatics.

GENOMICS, PROTEOMICS AND BIOREMEDIATION

All living things, including microorganisms, have a chemical called DNA (deoxyribo-nucleic acid) that contains information used in the organism to build and maintain cell biomass and to reproduce itself. The DNA molecule is made up of four chemical building blocks (bases): adenosine (A), thymine (T), cytosine (C) and guanine (G). In microorganisms millions of these bases form long strand that pair together (A with T and C with G) in a twisted zipper-like structure known as a "double helix".

Genomics is the study of the complete set of genetic information—all the DNA in an organism. This is known as its genome. Genomes range in size: the smallest known bacterial genome containing about 600,000 base pairs and the human genome has some three billion. (The size of a genome is designated in millions of base pairs of the megabytes, abbreviated Mb.) Typically genes are segments of DNA that contain instructions on how to make the proteins that code for structural and catalytic functions. Combination of genes, often interacting with environmental factors, ultimately determine the physical characteristics of an organism.

Genomes Sequencing

How are the genomes sequenced? Microbial genomes are first broken into shorter pieces. Each short piece is used as a template to generate a set of fragments that differ in length from each other by a single base. The last base is labeled with a fluorescent dye specific to each of the four base types. The fragments in a set are separated by gel electrophoresis. The final base at the end of each fragment is identified using laser induced fluorescence, which discriminates among the different labeled bases. This process recreates the original sequence of bases (A, T, C and G) for each short piece generated in the first step.

Automated sequencers analyze the resulting electropherograms, and the output is a four colour chromatogram showing peas that represent each of the four DNA bases. After the bases are "read", computers are used to assemble the short sequences (in blocks of about 500 or more bases each, called the read length) into long continuous stretches that are analyzed for errors, gene-coding regions and other characteristics. To generate a high quality sequences, additional sequencing is needed to close gaps, reduce ambiguities and allow for only a single error every 10,000 bases. By the end of the process, the entire genome will have been sequenced the equivalent of 8 or 9 times. The finished sequence is submitted to major public sequence database, such as GenBAnk (http://www.ncbi.nih.gov).

Once the genome has been sequenced, portions that define features of biological importance must be identified and annotated. When the newly identified gene has a close relative already in a DNA database, gene finding is relatively straight forward. The genes tend to be simple, uninterrupted open reading frames (ORFs) that can be translated and compared with the database. However, the discovery of new genes without close relatives is more problematic. Scientists in the new discipline of bioinformatics are developing and applying computational tools and algorithms to help identify the functions of these previously unidentified genes. An accurate accounting and description of genes in microbial genome is essential for describing metabolic pathway and other aspects of whole-organism function.

GENOME SEQUENCING—BIOINFORMATICS

The major contribution of the bioinformatics in genome sequencing has been in:

- The development of automated sequencing technique that integrate the PCR or BAC based amplification, 2D gel electrophoresis and automated reading of nucleotides.
- Joining the sequences of smaller fragments (contigs) together to form a complete genome.
- The prediction of promoters and protein coding regions of the genome.

Polymerase Chain Reaction (PCR) or Bacterial Artificial Chromosome (BAC) based amplification techniques derive limited size fragments of a genome. The available fragment sequences suffer from nucleotide reading errors, very small repeats and very similar fragments that fit in two or more parts of a genome. Chimera is two different parts of the genome or artifacts caused by contamination that join end to end giving a artifactual fragment. Generating multiple copies of the fragments, aligning the fragments and using the majority voting at the same nucleotide positions solve the nucleotide reading error problem. Multiple experimented copies are needed to establish repeats and chimeras. Chimeras and repeats are removed before the assembly of the genome fragments. The joining of the fragments is modeled as a mathematical weighted graph where nodes are fragments and the weights of edges are the number of overlapping nucleotides and the fragments are joined based upon maximum overlap using a greedy algorithm (Mount, 2000; Waterman, 1995). In a greedy algorithm, most nodes having maximum (or minimum) scores are collapsed first. To join contigs, the fragments with larger nucleotide sequence overlap are joined first.

Automated Identification of Genes

After the contigs are joined, the next issue is to identify the protein coding regions or ORFs (Open Reading Frames) in the genomes. The identification of ORFs can be done in three ways:

- Using Hidden Markov Mode (HMM) based techniques such as GLIMMER (Delcher et al., 1999) and genemark
- By searching the known database of genes such as GenBank ftp:// ftp.ncbi.nih.gov/genbank to identify genes and
- The use of algorithms based on decision trees that identify start codons (Suzek et al., 2001) and stop codons of the coding regions.

HMM-based techniques develop multiple probabilistic state machines each capable of identifying an ORF. Each machine predicts the next nucleotide character using a state transition with maximum probability and matches the predicted nucleotide character with the current nucleotide character in the actual sequence. Statistical training using known sample sequences derives the probability of state transition. In the case of microbial genome, HMM-based software such as GLIMM has provided 95-97% accuracy.

Identifying Gene Function: Searching and Alignment

After identifying the ORFs (Open Reading Frames) the next step is to annotate the gene with proper structure and function. The function of the gene has been identified using popular sequence search and pairwise gene alignment techniques. The four most popular algorithms used in functional annotation of the genes are:

- BLAST (Altschul et al., 1990) and its variations (Altschul et al., 1997)
- Dynamic programming technique—Smith-Waterman alignment (Waterman, 1995) and its variations
- Indexing based scheme FASTA (Pearson and Lipman, 1988) and its variations, and
- BLOCKS (Henikoff et al., 1995) that uses multiple sequence alignment of conserved domains to identify motifs-characterizing patterns of proteins.

3D Structure Modeling

A protein may live under one or more low free-energy conformational states depending upon its interaction with other proteins. Under a stable conformational state certain regions of the protein are exposed for protein-protein or protein-DNA interactions. Since function is also dependent upon exposed active sites, protein function can be predicted by matching the 3D structure of an unknown protein with the 3D structure of known protein (Baker and Sali, 2001). However, 3D structure from X-Ray crystallography and NMR spectroscopy are limited. Thus there is a need for alternate mechanism to match genes. Generally there is close correspondence between gene sequence and 3D structure. In such cases sequence matching is sufficient for function annotation. However, many times multiple sequences map to the same 3D structure; the lack of matching of amino acid sequences does not exclude same 3D structure. In such cases matching 2D structure (Kitagawa et al., 2001; Knaebel and Crawford, 1995)—patterns of alpha helix and beta sheets—and matching 3D structure is needed to verify the function of the newly sequenced protein (Whisstock and Lesk, 2003).

Pairwise Genome Comparison

After the identification of gene-functions, a natural step is to perform pairwise genome comparisons. Pairwise genome comparison of a genome against itself provides the details of paralogous genes—duplicated genes that have similar sequence with some variation in function. Pairwise genome comparisons of a genome against other genomes have been used to identify a wealth of information such as orthologous genes—functionally equivalent genes diverged in two genomes due to specification; different types of gene-groups—adjacent genes that are constrained to occur in close proximity due to their involvement in some common higher function; lateral gene-transfer—gene transfer from a microorganism that is evolutionary distant; gene-fusion/gene-fission; gene group duplication; gene-duplication and difference analysis to identify gene specific to a group of genomes such as pathogens, and conserved genes (Bansal, 1999; 2002).

Reconstructing Metabolic Pathways

Identification of gene functionality has started a new level of bioinformatics research: automated reconstruction and comparison of pathways of newly sequence organisms (Bansal, 2001; Bono et al., 1998; Kelley et al., 2003; Ogata et al., 1999; Papin et al., 2002; Schilling et al., 2002). There have been many efforts and approaches related to pathway reconstruction. The three major approaches can be classified as:

*   Global network of reactions catalyzed by enzymes;
*   Network of gene-groups connected through the reactions catalyzed by enzymes embedded in the gene groups; and
*   Global modeling of chemical reactions in the microbial cells.

Current metabolic pathway techniques are limited by the available gene-functions from wet laboratories. Another issue is that the identification of metabolic pathways is not sufficient unless the reaction rates are known. While there have been recent approaches to model the reaction rate of metabolic pathway (Schuster, 1999), the complete picture cannot be verified largely due to unavailability of gene-functions from wetlabs.

PROTEOMICS

Proteomics is a new branch of science dedicated to studying all the proteins expressed by a cell and how these proteins change under different growth conditions. For example, scientists are trying to understand the protein expression pattern of the organism. The proteins mainly responsible for the conversion are located on the outer membrane of the cell. Protein expression can be measured by isolating the proteins from cells and determining their relative abundance using protein separation and detection methods such as two-dimensional gel electrophoresis.

Mass spectrometry-based techniques are also being used to identify the full complement of cellular proteins, with the ultimate goal of determining how the cell localizes different proteins on the cell surface when conditions change from aerobic to anaerobic respiration. The entire complement of proteins associated with bacterial outer membrane vesicles (MVs) has been determined by a new technique that involves the use of both high-resolution separation and high-mass accuracy and sensitivity. Fourier Transform Ion Cyclotran Resonance (FTICR) mass spectrometry, MVs are unique to gram negative bacteria and are constantly being released from the cell surface during bacterial growth. During their release, MVs trap some of the underlying periplasm that contains various enzymes. MVs explain how bacteria protect enzymes that are secreted extracellularly and also

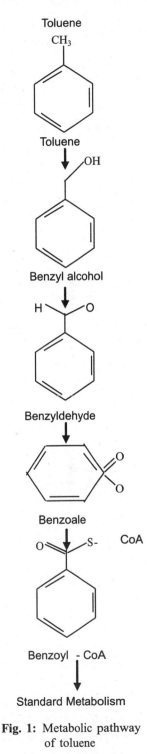

Fig. 1: Metabolic pathway of toluene

the method by which they deliver lethal enzymes into other bacteria as a means of predation, and even to eukaryotic cells early in pathogenesis. Identifying the entire protein complement of MVs can lead to predictions of their impact on the environments in which they are found.

Phenotype Similarity and Automated Pathway Comparisons

The researchers have taken the next level of study to compare the similar pathways to understand the effect of insertion and detection of genes in various microorganisms and to understand the evolution of pathway level (Bansal and Woolverton, 2003). To compare two pathways, the genes in the pathway are aligned as follows:

• Two pathways match completely if every protein in the first pathway (or a gene-group within a pathway) has a corresponding homologous gene in another pathway (or the gene-group within the pathway).

There is a gap if a homologous gene is deleted (inserted) and there is a mismatch if the corresponding homologous genes have a low similarity score. Based upon this modeling, comparison of H.Pylori and Yeast has shown many similar pathways. More importantly, quantification mechanism has been found to compare two pathways.

The terms 'proteomics' and 'proteome' were introduced in 1995 which is a key post-genomic feature that emerged from the growth of large and complex genome sequencing datasets. Proteomic analysis is particularly vital because the observed phenotype is a direct result of the Action of the protein rather than the genome sequence. Traditionally, this technology is based on highly efficient methods of separation using two-dimensional polyacrylamide gel electrophoresis (2-DE) and modern tools of bioinformatics in conjunction with mass spectrometry (MS) (Tatusov, 1996). In bioremediation, the proteome of the membrane proteins is of high interest, specifically in PAH biodegradation.

Sudip K. Samanta et al., 2001 discussed the problems of PAH pollution and PAH degradation and relevant bioremediation. The microbial degradation of PAH is the action of dioxygenase, which incorporates atoms of oxygen at the carbon atoms of a benzene ring of a PAH resulting in the formation of cis-dihydrodiol (Kanaly and Harayama, 2000), which undergoes rearomatization by dehydrogenases to form dihydroxylated intermediates. Dihydroxyated intermediates subsequently undergo ring cleanage and form TCA cycle intermediates (Sabate et al., 1999). A large number of naphthalene-degrading microorganisms—*Alcaligenes denitrificans, mycobacterium* sp., *Pseudomonas putida, P. fluorescens, P. paucimobilis, P. vesicularis, P. cepacia, P. testosteroni, Rhodococcus* sp., *corynebacterium venale, bacillus cereus, moraxella* sp., *streptomyces* sp., *vibrio* sp. and *cyclotrophicus* sp.—has been isolated and examined for mineralization (Hedlund and James, 2001; Samanta et al., 2001).

During the past decade, a variety of microorganisms has been isolated and characterized for the ability to degrade different PAHs and new pathways for PAH degradation elucidated. Further research is needed to explore the microbial interactions within PAH-degrading consortia, the regulatory mechanisms of various ring structured PAH biodegradation as well as the co-metabolic biodegradation of PAHs. The new approach of advancements in molecular biology can aid in the detection of PAH-degrading organisms from environmental samples (Watanabe, 2001; Watanabe and Baker, 2000). DNA-DNA hybridization has been directly applied to detect and monitor the crucial populations recovered from the environment (Sayler et al., 1985 and Guo et al.,1997). Laurie and Jones (2000) detected two distinct PAH catabolic genotypes from aromatic hydrocarbon-contaminated soil using quantitative competitive PCR (QC-PCR). A soil-derived consortium capable of rapidly mineralizing benzo[a]pyrene was analysed using Denaturing Gradient Gel Electrophoresis (DGGE) profiling of PCR-amplified 165 rDNA fragments (Laurie and Jones, 2000). This analysis detected 165 rDNA sequence types that represented organisms closely related to known high molecular weight PAH degrading bacteria. Recently, many PAH degrading bacteria have been isolated from geographically diverse sites using the 165 rRNA sequence technique (Widada et al., 2002).

Discovery of Novel Catabolic Genes Involved in Xenobiotic Degradation

There are two different approaches to investigate the diversity of catabolic genes in environmental samples: culture-dependent and culture-independent methods. In culture-dependent methods, bacteria are isolated from environmental samples with culture medium. Nucleic acid is then extracted from the bacterial culture. In contrast, culture-independent methods employ direct extraction of nucleic acids from environmental samples (Okuta et al., 1998; Watanabe et al., 1998). The description of catabolic gene diversity by culture-independent molecular biological methods often involves the amplification of DNA or cDNA from RNA extracted from environmental samples by PCR, and the subsequent analysis of the diversity of amplified molecules (community fingerprinting). Alternatively, the amplified products may be cloned and sequenced to identify and enumerate bacterial species present in the sample.

To confirm that the proper gene has been PCR-amplified, it is necessary to sequence the product, after which the resultant information can be used to reveal the diversity of the corresponding gene(s). Over the last few years, these molecular techniques have been systematically applied to the study of the diversity of aromatic-compound-degrading genes in environmental samples (Table 1).

Table 1: Molecular approaches for investigating the diversity and identification of catabolic genes involved in degradation of xenobiotics

Target gene	Molecular approach	Source	Reference
NahAc	RT-PCR with degenerate primers	Groundwater (culture- independent)	Wilson et al., 1999
phnAc, nahAc, and glutathione-S-transferase	PCR with several primers	Soil samples (culture-independent)	Lloyd-Jones et al., 1999
Phenol hydroxylate (LmPH)	PCR-DGGE with degenerate primers	Activated sludge (culture-dependent)	Watanabe et al., 1998
RHD	PCR with degenerate primers	Prestine- and aromatic hydrocarbon contaminated soils (culture-independent)	Yeates et al., 2000
PAH dioxygenase	PCR with several primers	PAH soil bacteria (culture-dependent)	Lloyd-Jones et al., 1999
NahAC	PCR with degenerate primers	Marine sediment bacteria (culture-dependent)	Hedlund et al., 1999
Nah	PCR with degenerate primers	Soil bacteria (culture-dependent)	Hamann et al., 1999
tfdC	PCR with degenerate primers	Soil bacteria (culture-dependent)	Cavalca et al., 1999
PAH dioxygenase and catechol dioxygenase	PCR with degenerate primers	Wastewater and soil bacteria (culture-dependent)	Meyer et al., 1999
phnAc, nahAc, PAH dioxygenase	PCR with several degenerate primers	River water, sediment and soil bacteria (culture-dependent)	Widada et al., 2002a
RHD	PCR-DGGE with degenerate primers	Thodococcus sp. strain (RHAI) (culture-dependent)	Kitagawa et al., 2001

RT: Reverse transcription, PCR: polymerase chain reaction, DGGE: denaturing gradient gel electrophoresis, RHD: ring hydroxylating dioxygenase, PAH: polycyclic aromatic hydrocarbon

Monitoring of Bioaugmented Microorganisms in Bioremediation

Because different methods for enumeration of microorganisms in environmental samples sometimes provide different results, the method used must be chosen in accordance with the purpose of the study. Not all detection methods provide quantitative data; some only indicate the presence of an organism and others only detect cells in a particular physiological state. Several molecular approaches have been developed to detect and quantify specific microorganisms (Table 2).

Table 2: Molecular approaches for detection and quantification of specific microorganisms in environmental samples

Identification method	Detection and quantification method	Cell type monitored
Fluorescent tags on rRNA probes	Microscopy Flow cytometry	Primary active cells
Lux and luc gene	Luminometry/scintillation counting	Active cells
	Cell extract luminescence	Total cells with translated luciferase
	Luminescent colonies	Curable luminescent cells
Gfp gene	Fluorescent colonies Microscopy Flow cytometry	Curable fluorescent cells Total cells, including starved
Specific DNA sequence	cPCR MPN-PCR, RLD-PCR	Total DNA (living and dead cells and free DNA)
	Slot/dot blot hybridization Colony hybridization	Culturable cells
Specific mRNA transcripting	Competitive RT-PCR	Catabolic activity of cells
	Slot/dot blot by hybridization	
Other marker genes (e.g., lacZY, gusA, xylE, and antibiotic resistance genes)	Plate counts colony hybridization	Culturable marked cells and indigenous cells with marker phenotype
	Quantitative PCR Slot/blot hybridization	Total DNA (living and dead cells and free DNA)

cPCR: Competitive PCR, MPN-PCR: most probable number PCR, RLD-PCR: replicative limiting dilution PCR
Source: J. Widada, H. Nojiri, and T. Omori 2002

In bioremediation, quantitative PCR has been used to monitor and to determine the concentration of some catabolic genes from bioaugmented bacteria in environmental samples (Table 3). Recently, quantitative competitive

RT-PCR has been used (Levesque et al., 1997) to quantify the mRNA of the tcbC of *Pseudomonas* sp. strain P51.

Table 3: PCR detection and quantification of introduced bacteria in bioremediation of xenobiotics

Bacteria	Target gene	Detection and quantification method	Reference
Desulfitobacterium frappieri strain PCP-1 (pentachlorophenol-degrader)	16 rRNA	Nested PCR	Levesque et al., 1997
Mycobacterium chlorophenolicum strain PCP-1 (pentachlorophenol-degrader)	16 rRNA	MPN-PCR	Elsas van et al., 1997
Sphingomonas chlorophenolica (pentachlorophenol-degrader)	16 rRNA	Competitive PCR	Elsas van et al., 1998
Pseudomonas sp. strain B13 (chloroaromatic-degrader)	16 rRNA	Competitive PCR	Leser et al., 1995
Pseudomonas putida strain mx (toluene-degrader)	xylE	Competitive PCR	Hallier-Soulier et al., 1996
P. putida strain G7 (naphthalene-degrader)	nahAc	PCR-southern blot	Herrick et al., 1993
P. putida strain mt2 (toluene-degrader)	xylM	Multiplex PCR-Southern blot	Knaebel and Crawford, 1995
Alcaligenes eutropus strain JMP 134 [2,4-dichlorophenoxyacetic acid (2,4-D)-degrader]	tfdA		
P. putida ATCC 11172 (phenol-degrader)	dmpN	PCR-RT-PCR	Selvaratnam et al., 1997
Pseudomonas sp. strain P51 (trichlorobenzene-degrader)	tbcAa, tbcC	PCR	Tchelet et al., 1999
Pseudomonas sp. strain P51 (trichlorobenzene-degrader)	tbcC	Competitive DT-PCR	Meckenstock et al., 1998
Pseudomonas resinovorans strain CA 10 (carbozole-and dibenzo-dioxin-degrader)	carAa	Real-time competitive PCR	Widada et al., 2001, 2002b

Source: J. Widada, H. Nojiri, and T. Omori 2002

Derivation of regulatory mechanism and pathways

The genomics and proteomics research front has progressively moved from metabolic pathway reconstruction to the identification of signaling pathways and promoter analysis to identify transcription factors for protein-DNA interactions. There are four major approaches to study protein-DNA interactions:

(i) Micro-array analysis of gene-expressions under different stress conditions of cells;

(ii) Statistical analysis of promoter regions of orthologous genes (functionally equivalent genes in different organisms identified as best homologs);

(iii) Global analysis of frequency patterns of divers in the intergenic region—promoter region occurring between adjacent protein coding regions of a genome; and

(iv) Biochemical modeling at the atomic band level to understand how a protein will bind to nucleotides.

The microarray analysis technique is based upon experimental data, and other three approaches are based on mathematical modeling and sequence analysis. Microarray analysis (Sung-Keun et al., 2004) measures the relative change in the gene-expressions for a stressed (or a stimulated) cell and a change in cellular expression pattern differentiation, cellular cycle, tissue remodeling, sporulating etc.

MICRO-ARRAY TECHNIQUES

Gene Detection in Biodegradation in Microbial Communities

Sung-Keun Rhee (2004) and team of scientists (Atlas, 1981), at Environmental Science Division, Oak Ridge National Laboratory, studied the detection of genes involved in biodegradation and biotransformation in microbial communities by using 50 mer oligonucleotide Microarrays. Research findings highlight that to effectively monitor biodegrading populations, a comprehensive 50 mer oligonucleotide microarray was developed based on most of the 2,402 known genes and pathways involved in biodegradation and metal resistance. This array contained 1,662 unique and group-specific probes with <85% similarity to their non-target sequences. Based on artificial probes, results showed that under hybridization conditions of 50°C and 50% formamide, the 50 mer microarray hybridization can differentiate sequences having < 88% similarity. Specificity tests with representative pure cultures indicated that the designated probes on the arrays appeared to be specific to their corresponding target genes. The detection limit was 5 to 10 ng of genomic DNA in the absence of background DNA and 50 to 100 ng of pure-culture genomic DNA in the presence of background DNA or 1.3×10^7 cells in the presence of background RNA. Strong linear relationships between the signal intensity and the target DNA and RNA were observed ($r^2 = 0.95$ to 0.99). Application of this type of microarray to analyse naphthalene-amended enrichment and soil microcosms demonstrated that microflora changed differently depending on the incubation conditions, while the naphthalene-degrading genes from rhodococcus type microorganisms were dominant in naphthalene-degrading enrichments. The genes involved in naphthalene (and polyaromatic hydrocarbon and nitrotoluene) degradation from gram negative microorganisms such as Palstonia, Comamonas and Burkholderia were most abundant in the soil microcosms. In contrast to general conceptions, naphthalene degrading genes from pseudomonas were not detected, although pseudomonas is widely known as a model microorganism for studying naphthalene degradation. The real-time PCR analysis with four representative genes showed that the microarray-based quantification was very consistent with real time PCR (r^2 = 0.74). In addition, application of the array to both polyaromatic-hydrocarbon and benzene-toluene-ethylbenzene-xylene-contaminated and uncontaminated soils indicated that the developed microarrays appeared to be useful for profiling difference in microbial community structures. The results indicate that this technology has potential as a specific, sensitive and quantitative tool in revealing a comprehensive picture of the compositions of biodegradation genes and the microbial community in contaminated environments although more work is needed to improve detection sensitivity.

The transformation of environmental contaminants is a complex process that is influenced by the nature and amount of contaminant present, the structure and dynamics of the indigenous microbial community, and the interplay of geochemical and biological factors of contaminated sites (Gibson and Sayler, 1992; Leahy and Colwell, 1990; Golyshin et al., 2003). A better understanding of the processes inherent in natural bioremediation requires, in part, a better understanding of microbial ecology. However, conventional molecular methods (PCR-based technologies, such as gene cloning, terminal-restriction fragment length polymorphism, and denaturing gradient gel electrophoresis and in situ hybridization) for assessing microbial community structure and activities are labour-intensive. Rapid, quantitative and cost effective tools that can be operated in field scale heterogeneous environments are needed for measuring and evaluation bioremediation strategies.

DNA Micro-array in Bioremediation

Even with the complete genome sequences of microorganisms with the potential for bioremediation (Tiedje, 2002; Seshadri et al., 2005), studies are not accelerating in a rapid manner with the completed genome sequences. It is possible to analyse the expression of all genes in each genome under various environmental conditions using whole-genome DNA microarrays (Schut et al., 2003; Lovley, 2003; Dennis et al., 2003). Such genome-wide expression analysis provides important data for identifying regulatory circuits in these organisms (Rabus et al., 2005; Muffler et al., 2002; Schut et al., 2001). In the past DNA microarrays have been used to evaluate the physiology of pure environmental cultures (Rhee et al., 2004) and to monitor the catabolic gene expression profile in mixed microbial communities (Ye et al., 2000). More than 100 genes were found to be affected by oxygen-limiting conditions when a DNA microarray was used to study changes in mRNA expression levels in Bacillus subtilis grown under anaerobic conditions (Denef et al., 2003).

Sensitivity may often be a part of the problem in PCR-based cDNA microarrays, since only genes from populations contributing to more than 5% of the community DNA can be detected. Several parameters were evaluated to validate the sensitivity of spotted olifonucleotide for bacterial functional genomics (Cho and Tiedje, 2002). Optimal parameters were found to be 5'-C_6 amino-modified 7-mers printed on CMT-GAPS II substrates at a 40 μm concentration combined with the use of tyramide signal amplification labelling. Based on most of the known genes and pathways involved in biodegradation and metal resistance a comprehensive 50-mer-based oligoneucleotide microarray was developed for effective monitoring of biograding population (Kuhner et al., 2005). This type of DNA microarray was effectively used to analyse naphthalene-amended enrichment and soil microcosms demonstrated that microflora changed differentially depending on the incubation conditions (Kitagawa et al., 2001). A global gene expression analysis revealed the co-regulation of several thus-far unknown genes during the degradation of alkylbenzene. Besides this, DNA microarray have been used to determine bacterial species, in quantitative applications of stress gene analysis of microbial genomes and in genome-wide transcriptional profiles (Gao et al., 2004).

The microarray is a powerful genome technology that is widely used to study biological processes. Although microarray technology has been used successfully to analyse global gene expression in pure cultures (Liu et al., 2003; Lockhart et al., 1996; Schena et al., 1996; Thompson et al., 2002; Wodicka et al., 1997; Zhou, 2003), adapting microarray hybridization for use in environmental studies present great challenges in terms of specificity, sensitivity and quantitation (Zhou and Thompson, 2002; Call et al., 2001). Although microarray-based genomic technology has attracted tremendous interest among microbial ecologists, it has only recently been extended to study microbial communities in the environment (Guschin et al., 1997; Loy et al., 2002; Small et al., 2001; Tiquia et al., 2004; Valinsky et al., 2002; Wu et al., 2001; Michael et al., 1999).

A system of cluster analysis for genome-wide expression data from DNA microarray hybridization is described that uses standard statistical algorithms to arrange genes according to similarity in pattern of gene expression. The output is displayed graphically, conveying the clustering and the underlying expression data simultaneously in a form intuitive for biologists. It has been found in the budding yeast, Saccharomyces Cerevisiae, that clustering gene expression data groups together and find a similar tendency in human data. Thus patterns seen in genome-wide expression experiments can be interpreted as an indication of the status of cellular processes. Also, co-expression of genes of known function with poorly characterized or novel genes may provide a simple means of gaining leads to the functions of many genes for which information is not available currently.

The rapid advance of genome scale sequencing has driven the development of method to exploit this information by characterizing biological processes in new ways. The knowledge of the coding sequences of virtually every gene in an organism, for instance, invites development of technology to study the expression of all of them at once, because the study of gene expression of genes one by one has already provided a wealth of biological insight. To this end, a variety of techniques have been evolved to monitor rapidly and efficiently, transcript abundance for all of an organism's genea (Michael et al., 1999). Within the mass of numbers produced by these techniques, which amount to hundreds of data points for thousands or tens of thousands of genes it is an immense amount of biological information.

STRUCTURE DETERMINATION

Meta Router is a system for maintaining heterogeneous information related to biodegradation in a framework that allows its administration and mining (application of methods for extracting new data). It is an application intended for laboratories working in this area which need to maintain public and private data, linked internally and with external databases, and to extract new information from it.

Working with "States" (sets of compounds) attempts to stimulate an environment with a set of pollutants where a given reaction, carried out by a given bacteria, can modify one of the pollutants but not the other which "moves" the system to another "state" (another set of compounds) where another bacteria can act, etc. One would wonder which enzymes are needed to end up a state and which are the bacteria that have these enzymes.

The properties of compounds required for compound administration include:

- Main technical characteristics
- Compound queries
- Enzyme queries
- Reaction queries
- Path finder

The data to be obtained in each of the above are described herewith.

Main Technical Characteristics

Initial Information:

- 740 organic compounds (2,167 synonyms)
- 820 reactions
- 502 enzymes
- 253 organisms

- Name
- EC code
- Organisms
- Database sequence identifiers
- Links to other databases

- Name
- Synonyms
- SMILES Code
- Molecular weight
- User defined properties
  - density
  - evaporation rate
  - melting point
  - boiling point
  - water solubility
- Formula
- Image of the chemical structure
- Structure including PDB format
- Links to other databases

Compound Queries

- Name
- Synonyms
- Part of their name
- Part of their smiles code
  - C=O >> comps containing carbonyl group
  - CCCCC >> comps with five or more linear saturated carbons
- A range of molecular weight
- A range of values of associated properties (solubility, density, etc.)

- Name (and synonyms)
- Smiles code
- Formula
- Image of the chemical structure
- 3D structure in PDB format
- Molecular weight
- List of properties and associated values
- UMBBD code
- "Find degradative pathway"

Enzyme Queries

Selection of compounds by:
- Enzyme name
  - part of name
- EC code
  - part of code
- Organisms
- Combination of some of these (i.e. EC= 1 for oxidoreductases, organism=pseudomonas)

Information shown:
- Enzyme name
- UMBBD code
- EC code
- Organisms
- Associated reactions
- Links to each database

Reaction Queries

Selection of compounds by:
- Substrate(s)
  - synonyms
  - partial names
- Product(s)
  - synonyms
  - partial names
- Enzyme(s)
- Organism(s)
More than one substrate, product, on enzyme can be selected with AND/OR (at the bottom of the list)

Information shown :
- Chemical structures
  - substrate
  - products
- Name of enzyme
- UMBBD code of reaction
- Links to databases
  - compounds
  - enzymes
  - UMBBD page

Path Finder

- Localization of pathways from an initial set of compounds to a final one and/or to the standard metabolism.
- Selection of the pathways by length, organisms where the enzymes are present and characteristics of the implicated chemical compounds.
- Representation of the pathways with compound name, compound image, synonyms, formula, molecular weight, SMILES code and enzyme; hyperlinked to the corresponding information for compounds, enzymes and reactions.
- Colouring pathways according with compound properties and/or enzymatic classes.
- The University of Minnesota Biocatalysis/Biodegradation Database, UMBBD, (http://umbbd.ahc.umn.edu) is the largest resource of information about biodegradation on the Internet.
- ENZYME is a repository of information on enzymes (nomenclature, sequence, etc.) (http://www.expasy.ch/enzyme/).
- SMILES is a system for coding chemical compounds as linear strings of ASCII characters. It was developed by Daylight Chemical Information Systems, Inc. (http://www.daylight.com/smiles/f-smiles.html).
- SRS is a system for indexing, connecting and querying molecular biology databases (http://srs.ebi.ac.uk/). Although the system belongs to Lion Bioscience (http://www.1ionbioscience.con11), they maintain a free academic version.
- SQL (Structured Query Language) was developed by IBM as a standard language for interrogating relational databases implemented in most commercial and free database systems with little differences. The variant used in MetaRouter is that implemented in PostgreSQL (http://www.postgresql.org).

CONCLUSION

Bioinformatics, despite being a young field, has helped both fundamental microbiology and biotechnology through the development of algorithms, tools and discoveries refining the abstract model of microbial cell functioning. The major impact of the bioinformatics has been in automating the microbial genome sequencing, the development of integrated databases over the internet, and analysis of genomes to understand gene and genome function. The current front has moved to the identification of regulation pathways identification of protein-protein interactions, protein-DNA interactions, protein-RNA interactions, and simulations of metabolic reactions to study the effect of reaction rates and the analysis of experimental data available from microarray data to study the correlation between the gene-expression and stress conditions. Bioinformatic techniques are critically dependent upon the knowledge derived from wet laboratories and the available computational algorithms and tools. Both the resources have limited capability of handling a vast amount of data to interpret genomics and proteomics with so many unknowns. The progress in bioinformatics and wet lab techniques has to remain interdependent and focused complementing each other for their own progress and for the progress of biotechnology.

The application of molecular biology-based techniques in bioremediation is being increasingly used and has provided useful information for improving of bioremediation strategies and assessing the impact of bioremediation treatments on ecosystems. Recent developments in molecular biology techniques also provide rapid, sensitive and accurate methods by analyzing bacteria and their catabolic genes in the environment.

FUTURE: SCIENTIFIC UNDERSTANDING

Microbes found in the contaminated subsurface and other environments often have the metabolic capability to degrade or otherwise transform contaminants. Microbes can directly or indirectly, through their influence on sediment geochemistry, provide a potential cost-effective bioremediation strategy to immobilize contaminants. Research is needed to provide useful information to decision makers on whether remediation is necessary and practical, give an accurate prediction of contaminant mobility, and suggest bioremediation strategies. A biotreatment technique that works well at one site may perform partly at another site because of lack of understanding of the unique interactions between the microbial community and geochemistry in a particular ecosystem. Mostly current research focuses on understanding single microbial species having potential for environmental remediation and stabilization in laboratory-based cultivation which, however, differs in the contaminated environment in the field. Therefore, the understanding needs the following:

- What is the makeup of microbial communities and the need to learn who is there, their physiological states, their individual contributions, how they relate to each other and the environment, and how metagenomic DNA sequence can be used to predict the functions, behaviour and evolutionary trajectory of microbial communities.
- How do microbes identify, access and modify their local geochemical environments to gain energy and nutrients and meet other metabolic requirements?
- What physiochemical environmental interactions control the dynamic makeup, structure and function of microbial communities in the subsurface, and what is the resultant impact on contaminant transformation?
- How and why do contaminants impact microbial communities and what are potential indicators of these impacts?
- How do molecular mechanistic processes to microbes and communities relate to macroscopic behaviour in field environments?

Bioremediation Challenges and Bioinformatics Data

Research Challenges Bioremediation	Scale and Complexity Bioinformatics
Analysis of:	*Needed Research on:*
• Microbial communities and their metabolic activities that impact the fate and transport of contaminants • Geochemical changes in subsurface environments due to microbial or chemical activity • Microbial processes in heterogeneous environments	• Hundred of different sites, millions of genes, thousands of unique species and functions • Functional analysis of potentially thousands of enzymes involved in microbe-mineral interactions, hundreds of regulatory processes and interactions; spatially resolved community formation, structure; influence on contaminant fate • Models at the molecular, cellular and community level incorporating signaling, sensing, metabolism, transport biofilm, cell-mineral interactions, incorporated into macromodels for fate and transport.

Towards a deep understanding of bioremediation, new techniques in molecular biology—particularly genetic engineering, transcriptomics, proteomics and interactions after remarkable advances—promise as tools to study the mechanisms involved in regulation of mineralization pathways. The application of these techniques are still in their infancy, but the amount of data that is continuously being generated by today's genomics and proteomics technocrats needs to be organised in a stepwise manner within informative databases. The strategies need to be refined in which transcriptomics and proteomics data are combining together in order to understand the mineralization process in a meaningful way. These techniques show great promise in their ability to predict organisms' metabolism in contaminated environments and to predict the microbial-assisted attenuation of contaminants to accelerate bioremediation.

Sites: Bioremediation

UM-BBD: University of Minesota Biocatalysis/Biodegradation database
http://www.umbbd.ahc.umn.edu/index.html

KEGG: Kyoto Encyclopedia of Genes and Genomes
http://www.genome.ad.jp/kegg/kegg.html

Boehringer Mannhein Biochemical Pathways on the ExPASy server, Switzerland
http://www.expasy.org/cgi-bin/search-biochem-index

Enzyme and Metabolic Pathway (EMP) database at Argonne National Laboratories
http://emp.mcs.anl.gov

International Society for the Study of Xenobiotics
http://www.issx.org

Biopathways Consortium
http://www.biopathways.org

BioCyc: Knowledge Library of Pathway/Genome Databases
http://biocyc.org/

PathDB: Metabolic Pathways Database at NCGR
http://www.ncgr.org/pathdb/

Metabolic Pathway Minimaps at Trinity College, Dublin, Ireland
http://www.tcd.ie/Biochemistry/IUBMB-Nicholson/

Yeast Genome Pathways at MIPS, Germany
http://www.mips.biochem.mpg.de/proj/yeast/pathways/

REFERENCES

Allison, D.G., Ruiz, B., San-Jose, C., Jaspe, A. and Gilbert, P. (1998). Analysis of biofilm polymers of Pseudomonas fluorescens B52 attached to glass and stainless steel coupons. *In:* Abstracts of the General Meeting of the American Society for Microbiology, Atlanta, Georgia, **98:** 325.

Altschul, S.F., Gish, W., Miller, W., Myers, E.W. and Lipman, D.J. (1990). Basic alignment search tools. *Journal of Molecular Biology*, **215:** 403-410.

Altschul, S.F., Madden, T.L., Schaffer, A.A., Zhang, J., Zhang, Z., Miller, W. and Lipman, D.J. (1997). Gapped BLAST and PSI-BLAST: A new generation of protein database search programs. *Nucleic Acids Research*, **15(17):** 3389-3402.

Alves, R., Chaleil, R.A.G. and Sternberg, M.J.E. (2002). Evolution of enzymes in metabolism: A network perspective. *J. Mol. Biol.*, **320:** 751-770.

Atlas, R.M. (1981). Microbial degradation of petroleum hydrocarbons: An environmental perspective. *Microbiol. Rev.*, **7:** 285-292.

Baker, D. and Sali, A. (2001). Protein structure prediction and structural genomics. *Science*, **294:** 93-96.

Bansal, A.K. (2001). Integrating co-regulated gene-groups and pair-wise genome comparisons to automate reconstruction of microbial pathways. IEEE International Symposium on Bioinformatics and Biomedical Engineering, Washington, 209-216.

Bansal, A.K. (1999). An automated comparative analysis of seventeen complete microbial genomes. *Bioinformatics*, **15(11):** 900-908.

Bansal, A.K. and Meyer, T.E. (2002). Evolutionary analysis by whole genome comparisons. *Journal of Bacteriology*, **184(8):** 2260-2272.

Bansal, A.K. and Woolverton, C. (2003). Applying automatically derived gene-groups to automatically predict and refine microbial pathways. *IEEE Transactions of Knowledge and Data Engineering*, **15(4):** 883-894.

Bono, H., Ogata, H., Goto, S. and Kanehisa, M. (1998). Reconstruction of amino acid biosynthesis pathways from the complete genome sequence. *Cenome Research*, **8(3)**: 203-210.

Call, D., Chandler, D. and Brockman, F. (2001). Fabrication of DNA microarrays using unmodified oligonucleotide probes. *BioTechniques*, **30**: 368-379.

Cavalca, L., Hartmann, A., Rouard, N. and Soulas, G. (1999). Diversity of tfdC genes: distribution and polymorphism among 2,4-dichlorohenoxyacetic acid degrading soil bacteria. *FEMS Microbiol. Ecol.*, **29**: 45-58.

Cho, J.C. and Tiedje, J.M. (2002). Quantitative detection of microbial genes by using DNA microarrays. *Appl. Environ. Microbial.*, **68**: 1425-1430.

Delcher, A.L., Harmon, D., Kasif, S., White, O. and Salzberg, S.L. (1999). Improved microbial gene identification with GLIMMER. *Nucleic Acids Research*, **27(23)**: 4636-4641.

Denef, V.J., Park, J., Rodrigues, J.L. et al. (2003). Validation of a more sensitive method for using spotted oligonucleotide DNA microarrays for functional genomics studies on bacterial communities. *Environ. Microbial.*, **5**: 933-943.

Dennis, P., Edwards, E.A., Liss, S.N. et al. (2003). Monitoring gene expression in mixed microbial communities by using DNA microarrays. *Appl. Environ. Microbial.*, **69**: 769-778.

DeRisi, J.L., Iyer, V.R. and Brown, P.O. (1997). Exploring the metabolic and genetic control of gene expression on a genomic scale. *Science*, **278**: 680-686.

Ellis, L.B., Hou, B.K., Kang, W. and Wackett, L.P. (2003). The University of Minnesota Biocatalysis/Biodegradation Database: Post-genomic data mining. *Nucleic Acids Res.*, **31**: 262-265.

Elsas, J.D. van, Mantynen, V. and Wolters, A.C. (1997). Soil DNA extraction and assessment of the fate of Mycobacterium chlorophenolicum strain PCP-l in different soils by 16S ribosomal RNA gene sequence based most-probable-number PCR and immunofluorescence. *Bioi. Fertil. Soil*, **24**: 188-195.

Elsas, J.D. van, Rosado, A., Moore, A.C. and Karlson, V. (1998). Quantitative detection of Sphingomonas chlorophenoliza in soil via competitive polymerase chain reaction. *J. Appl. Microbiol.*, **85**: 463-471.

Fraser, H.B., Hirsh, A.E., Steinmetz, L.M., Scharfe, C. and Feldman, M.W. (2002). Evolutionary rate in the protein interaction network. *Science*, **296**: 750-752.

Gao, H., Wang, Y., Liu, X. et al. (2004). Global transcriptome analysis of the heat shock response of Shewanella oneidensis. *J. Bacterial*, **186**: 7796-7803.

Gibson, D.T. and Sayler, G.S. (1992). Scientific foundations of bioremediation: Current status and future needs. American Academy of Microbiology, Washington, D.C.

Golyshin, P.N., Martins Dos, Santos, V.A., Kaiser, O. et al. (2003). Genome sequence completed of Alcanivorax borkumensis: A hydrocarbon-degrading bacterium that plays a global role in oil removal from marine systems. *J. Biotechnol.*, **106**: 215-220.

Guo, C. et al. (1997). Hybridization analysis of microbial DNA from fuel oil-contaminated and noncontaminated soil. *Microbial. Ecol.*, **34**: 178-187.

Guschin, D.Y., Mobarry, B.K., Proudnikov, D., Stahl, D.A., Rittman, B.E. and Mitzabekov, A.D. (1997). Oligonucleotide microarrays as genosensors for determinative environmental studies in microbiology. *Appl. Environ. Microbiol.*, **63**: 2397-2402.

Hallier-soulier, S., Ducrocq, V., Mazure, N. and Truffaut, N. (1996) Detection and quantification of degradative genes in soils contaminated by toluene. *FEMS Microbiol. Ecol.*, **20**: 121-133.

Hamann, C., Hegemann, J. and Hildebrandt, A. (1999). Detection of polycyclic aromatic hydrocarbon degradation genes in different soil bacteria by polymerase chain reaction and DNA hybridization. *FEMS Microbiol. Lett.*, **173**: 255-263.

Hedlund, B.P., Geiselbrecht, A.D., Timothy, 18 and Staley, J.T. (1999). Polycyclic aromatic hydrocarbon degradation by a new marine bacterium, Neptunomonas napthovorans, sp. nov. *Appl. Environ. Microbiol.*, **65**: 251-259.

Henikoff, S., Henikoff, J.G., Alford, W.J. and Pietrokovski, S. (1995). Automated construction and graphical presentation of protein blocks from unaligned sequences. *Gene*, **163(2)**: GC17-26.

Herrick, J.B., Madsen, E.L., Batt, C.A. and Ghiorse, W.C. (1993). Polymerase chain reaction amplification of naphthalene-catabolic and 16S rRNA gene sequences from indigenous sediment bacteria. *Appl. Environ. Microbiol.*, **59**: 687-694.

Hochstrasser, D.F. (1998). Proteome in perspective. *Clin Chem Lab Med.* **36**: 825-836 [Cross Ref: ISI, Medline].

Ideker, T., Thorsson, V., Ranish, J.A., Christmas, R., Buhler, J., Eng, J.K., Bumgarner, R., Goodlett, D.R., Aebersold, R. and Hood, L. (2001). Integrated genomic and proteomic analyses of a systematically perturbed metabolic network. *Science*, **292**: 929-934.

Jansson, J.K. and Prosser, J.I. (1997). Quantification of the presence and activity of specific microorganisms in nature. *Mol. Biotechnol.*, **7**: 103-120.

Jeong, H., Mason, S.P., Barabási, A.L. and Oltvai, Z.N. (2001). Lethality and centrality in protein networks. *Nature*, **411**: 41-42.

Kanaly, R.A. and Harayama, S. (2000). Biodegradation of high-molecular-weight polycyclic aromatic hydrocarbons by bacteria. *J. Bacteriol.*, **182**: 2059-2067.

Kelley, B.P., Sharan, R., Karp, R.M., Sittler, T., Root, D.E., Stockwell, B.R. and Ideker, T. (2003). Conserved pathways within bacteria and yeast as revealed by global protein network alignment. *PNAS*, **100(20)**: 11394-11399.

Kitagawa, W., Suzuki, A., Hoaki, T., Masai, E. and Fukuda, M. (2001). Multiplicity of aromatic ring hydroxylation dioxygenase genes in a strong PCB degrader, Rhodococcus sp. strain RHA 1 demonstrated by denaturing gel electrophoresis. *Biosci. Biotechnol. Biochem.*, **65**: 1907-1911.

Knaebel, D.B. and Crawford, R.L. (1995). Extraction and purification of microbial DNA from petroleum-contaminated soils and detection of low numbers of toluene, octane and pesticide degraders by multiplex polymerase chain reaction and Southern analysis. *Mol. Ecol.*, **4**: 579-591.

Kuhner, S., Wohlbrand, L., Fritz, I. et al. (2005). Substrate-dependent regulation of anaerobic degradation pathways for toluene and ethylbenzene in a denitrifying bacterium, strain EbN1. *J. Bacterial*, **187**: 1493-1503.

Laurie, A.D. and Jones, G.L. (2000). Quantification of phnAc and nahAc in contaminated New Zealand soils by competitive PCR. *Appl. Environ. Microbiol.*, **66**: 1814-1817.

Liu, Y., Zhou, J.-Z., Omelchenko, M., Beliaev, A., Venkateswaran, A., Stair, J., Wu, L., Thompson, D.K., Xu, D., Rogozin, I.B., Gaidamakova, E.K., Zhai, M., Makarova, K.S., Koonin, E.V. and Daly, M.J. (2003). Transcriptome dynamics of Deinococcus radiodurans recovering from ionizing radiation. Proc. Natl. Acad. Sci. USA, **100**: 4191-4196.

Leahy, J.G. and Colwell, R.R. (1990). Microbial degradation of hydrocarbons in the environment. *Microbiol. Rev.*, **54**: 305-315.

Leser, T.D., Boye, M. and Hendriksen, N.B. (1995). Survival and activity of Pseudomonas sp. strain B 13(FR 1) in a marine microcosm determined by quantitative PCR and an rRNA-targeting probe and its effect on the indigenous bacterioplankton. *Appl. Environ. Microbiol.*, **61**: 1201-1207.

Levesque, M.J., La-Boissiere, S., Thomas, J.C., Beaudet, R. and Villemur, R. (1997). Rapid method for detecting Desulfitobacterium frappri strain PCP-1 in soil by the polymerase chain reaction. *Appl. Microbiol. Biotechnol.*, **47**: 719-725.

Lloyd-Jones, G., Laurie, A.D., Hunter, D.W.F. and Fraser, R. (1999). Analysis of catabolic genes for naphthalene and phenanthrene degradation in contaminated New Zealand Soils. *FEMS Microbiol Ecol.*, **29**: 69-79.

Lockhart, D.J., Dong, H., Byrne, M.C., Follettie, M.T., Gallo, M.V., Chee, M.S., Mittmann, M., Wang, C., Kobayashi, M., Horton, H. and Brown, E.L. (1996). Expression monitoring by hybridization to high-density oligonucleotide arrays. *Nat. Biotechnol.*, **14**: 1675-1680.

Lovley, D.R. (2003). Cleaning up with genomic: Applying molecular biology to bioremediation. *Nat. Rev. Microbial.*, **1**: 35-44.

Loy, A., Lehner, A., Lee, N., Adamczyk, J., Meier, H., Ernst, J., Schleifer, K.-H. and Wagner, M. (2002). Oligonucleotide microarray for 16S rRNA-based detections of all recognized lineages of sulfate-reducing prokaryotes in the environment. *Appl. Environ. Microbiol.*, **68**: 5064-5081.

Meckenstock, R., Steinle, P., van der Meer, J.R. and Snozzi, M. (1998). Quantification of bacterial mRNA involved in degradation of 1,2,4-trichlorobenzene by Pseudomonas sp. strain P51 from liquid culture and from river sediment by reverse transcriptase PCR (RT/PCR). *FEMS Microbiol. Lett.*, **167**: 123-129.

Meta Router: Bioinformatics in Bioremediation

Meyer, S., Moser, R., Neef, A., Stahl, U. and Kampfer, P. (1999). Differential detection of key enzymes of polyaromatic-hydrocarbon-degrading bacteria using PCR and gene probes. *Microbiol.*, **145**: 1731-1741.

Michael B. Eisen, Paul T. Spellman, Patrick O. Brown and David Botsein. Department of Genetics and Department of Biochemistry and Howard Hughes Medical Institute, Stanford University, School of Medicines, CA 94305, Proc. Natl. Acad. Sci. U.S. **96(19)**: 10943.

Mount, D.W. (2000). Bioinformatics: Sequence and Genome Analysis.Cold Spring Harbor Laboratory Press, New York.

Muffler, A., Bettermann, S., Haushalter, M. et al. (2002). Genome-wide transcription profiling of Corynebacterium glutamicum after heat shock and during growth on acetate and glucose. *J. Biotechnol.*, **98**: 255-268.

National Research Council (NRC) (1993). In situ bioremediation: When does it work? National Academies Press, Washington, D.C.

Ogata, H., Goto, S., Fujibuchi, W. and Kanehisa, M. (1999). Computation with the KEGG pathway database. *Biosystems*, **47**: 119-128.

Okuta, A., Ohnishi, K. and Harayama, S. (1998). PCR isolation of catechol 2,3-dioxygenase gene fragments from environmental samples and their assembly into functional genes. *Gene*, **212**: 221-228.

Papin, J.A., Price, N.D. and Palsson, B.Ø. (2002). Extreme pathway lengths and reaction participation in genome-scale metabolic networks. *Genome Research*, **12(12)**: 1889-1900.

Pearson, W.R. and Lipman, D.J. (1988). Improved tools for biological sequence comparison. *Proceedings National Academy of Science, USA*, **85(8)**: 2444-2448.

Rabus, R., Kube, M., Heider, J. et al. (2005). The genome sequence of an anaerobic aromatic-degrading denitrifying bacterium, strain EbN1. *Arch. Microbial.*, **183:** 27-36.

Rhee, S.K., Liu, X., Wu, L. et al. (2004). Detection of genes involved in biodegradation and biotransformation in microbial communities by using 50-mer oligonucleotide microarrays. *Appl. Environ. Microbial.*, **70:** 4303-4317.

Rison, S.G.C. and Thornton, J.M. (2002). Pathway evolution, structurally speaking. *Curr. Opin. Struct. Biol.*, **12:** 374-382.

Sabate, et al. (1999). Isolation and characterization of a 2-methylphenanthrene utilizing bacterium: Identification of ring cleavage metabolites. *Appl. Microbiol. Biotechnol.*, **52:** 704-712.

Samanta, S.K. et al. (2001). Efficiency of naphthalene and salicylate degradation by a recombinant Pseudomonas putida mutant strain defective in glucose metabolism. *Appl. Microbiol. Biotechnol.*, **55:** 627-631.

Samanta, S.K., Singh, O.V. and Jain, R.K. (2002). Polycyclic aromatic hydrocarbons: environmental pollution and bioremediation. Trends *Biotechnol*, **20:** 243-248.

Sayler, S. et al. (1985). Application of DNA-DNA colony hybridization to the detection of catabolic genotype in environmental samples. *Appl. Environ. Microbiol.*, **49:** 1295-1303.

Schena, M., Shalon, D., Heller, R., Chai, A., Brown, P.O. and Davis, R.W. (1996). Parallel human genome analysis: Microarray-based expression monitoring of 1000 genes. Proc. Natl. Acad. Sci. USA, **93:** 10614-10619.

Schilling, C.H., Covert, M.W., Famili, I., Church, G.M., Edwards, J.S. and Palsson, B.O. (2002). Genome-scale metabolic model of Helicobacter pylori 26695. *Journal of Bacteriology*, **184(6):** 4582-4593.

Schuster, S., Dandekar, T. and Fell, D.A. (1999). Detection of elementary flux modes in biochemical networks: A promising tool for pathway analysis and metabolic engineering. *Trends Biotechnology*, **17(2):** 53-60.

Schut, G.J., Brehm, S.D., Datta, S. et al. (2003). Whole-genome DNA microarray analysis of a hyperthermophile and an archaeon: Pyrococcus furiosus grown on carbohydrates or peptides. *J. Bacterial*, **185:** 3935-3947.

Schut, G.J., Zhou, J. and Adams, M.W. (2001). DNA microarray analysis of the hyperthermophilic archaeon Pyrococcus furiosus: Evidence for a new type of sulfur-reducing enzyme complex. *J. Bacterial*, **183:** 7027-7036.

Selvaratnam, S., Schoedel, B.A., McFarland, B.L. and Kulpa, C.F. (1997). Application of the polymerase chain reaction (PCR) and reverse transcriptase/PCR for determining the fate of phenol-degrading Pseudomonas putida A TCC 11172 in a bioaugmented sequencing batch reactor. *Appl. Microbiol. Biotechnol.*, **47:** 236-240.

Seshadri, R., Adrian, L., Fouts, D.E. et al. (2005). Genome sequence of the PCE-dechlorinating bacterium Dehalococcoides ethenogenes. *Science*, **307:** 105-108.

Small, J., Call, D.R., Brockman, F.J., Straub, T.M. and Chandler, D.P. (2001). Direct detection of 16S rRNA in soil extracts by using oligonucleotide microarrays. *Appl. Environ. Microbiol.*, **67:** 4708-4716.

Sung-Keun Rhee, Xueduan Liu, Liyonu Wu, Song C. Chong, Xiufeng Wan, and Jizhong Zhou (2004). Environmental Sciences Division, Oak Ridge National Laboratory, Oak Ridge, Tennessee TB 37831-6038. *Appl. Environ. Microbial*, **70(7):** 4303-4317.

Suzek, B.E., Ermolaeva, M.D., Schreiber, M., Salzberg, S.L. (2001). A probabilistic method for identifying start codons in bacterial genomes. *Bioinformatics*, **17(12):** 1123-1130.

Tatusov, R.L., Mushegian, M., Bork, P., Brown, N., Hayes, W.S., Borodovsky, M., Rudd, K.E. and Koonin, E.V. (1996). Metabolism and evolution of Haemophilius Influenzae deduced from a whole-genome comparison with *Escherichia Coli*. *Current Biology*, **6:** 279-291.

Tchelet, R., Meckenstock, R., Steinle, P. and van der Meer, J.R. (1999). Population dynamics of an introduced bacterium degrading chlorinated benzenes in a soil column and in sewage sludge. *Biodegradation*, **10:** 113-125.

Thompson, D.K., Beliaev, A.S., Giometti, C.S., Tollaksen, S.L., Khare, T., Lies, D.P., Nealson, K.H., Lim, H., Yates, J. III, Brandt, C.C., Tiedje, J.M. and Zhou, J.-Z. (2002). Transcriptional and proteomic analysis of a ferric uptake regulator (Fur) mutant of Shewanella oneidensis: Possible involvement of Fur in energy metabolism, transcriptional regulation, and oxidative stress. *Appl. Environ. Microbiol.*, **68:** 881-892.

Tiedje, J.M. (2002). Shewanella - The environmentally versatile genome. *Nat. Biotechnol.*, **20:** 1093-1094.

Tiquia, S.M., Chong, S.C., Fields, M.W. and Zhou, J. (2004). Oligonucleotide-based functional gene arrays for analysis of microbial communities in the environment. *In:* Kowalchuk, G.A., F.J. de Bruijn, I.M. Head, A.D. Lakkennans and J.D. van Elsas (eds), Molecular microbial ecology manual. Kluwer Academic Publishers, Dordrecht, The Netherlands.

Tiquia, S.M., Wu, L., Chong, S.C., Passovets, S., Xu, D., Xu, Y. and Zhou, J.-Z. (2004). Evaluation of 50-mer oligonucleotide arrays for detecting microbial populations in environmental samples. *BioTechniques*, **36:** 664-670, 672, 674-675.

URL : http://genomicsgtl.energy.gov, August 29, 2006.

Valinsky, L., Vedova, G.D., Scupham, A.J., Figueroa, A., Yin, B., Hartin, R.J., Chroback, M., Crowley, D.E., Jiang, T. and Borneman, J. (2002). Analysis of bacterial community composition by oligonucleotide fingerprinting of rRNA genes. *Appl. Environ. Microbiol.*, **68:** 3243-3250.

Waddell, Pl. and Kishino, H. (2000). Cluster inference methods and graphical models evaluated on NC160 microarray gene expression data. *Genome Informatics*, **11:** 129-140.

Wasinger, V.C., Cordwell, S.J., Cerpa-Poljak, A. et al. (1995). Progress with gene-product mapping of the Molecules: Mycoplasma genitallium Electrophoresis. **16:** 1090-1094 [Cross Ref: ISI, Medline].

Watanabe, K. (2001). Microorganisms relevant to bioremediation. *Curr. Opin. Biotechnol.*, **12:** 237-241.

Watanabe, K. and Baker, P.W. (2000). Environmentally relevant microorganisms. *J. Biosci. Bioeng.*, **89:** I-II.

Watanabe, K., Teramoto, M., Futamata, H. and Harayama, S. (1998). Molecular detection, isolation, and physiological characterization of functionally dominant phenol-degrading bacteria in activated sludge. *Appl. Environ. Microbiol.*, **64:** 4396-4402.

Waterman, M.S. (1995). Introduction to Computational Biology: Maps, Sequence, and Genomes. Chapman & Hall, London.

Whisstock, J.C. and Lesk, A.M. (2003). Prediction of protein function from protein sequence and structure. *Q. Rev. Biophysics*, **36(3):** 307-340.

Widada, J., Nijiri, H., Kasuga, K., Yoshida, T., Habe, H. and Omori, T. (2002a). Molecular and diversity of polycyclic aromatic hydrocarbon-degrading bacteria isolated from geographically diverse sites. *Appl. Microbiol. Biotechnol.*, **58**: 202-209.

Widada, J., Nojiri, H., Yoshida, T., Habe, H. and Omori, T. (2002b). Enhanced degradation of carbazole and 2,3-dichlorodibenzo-p dioxinin soils by Pseudomonas resinovorans strain CA1O. *Chemosphere.*

Wilson, M.S., Bakerman, C. and Madsen, E.L. (1999). In situ, real-time catabolic gene expression: extraction and characterization of naphthalene dioxygenase mRNA transcripts from groundwater. *Appl. Environ. Microbiol.*, **65**: 80-87.

Wodicka, L., Dong, H., Mittmann, M., Ho, M.H. and Lockhart, D.J. (1997). Genome-wide expression monitoring in Saccharomyces cerevisiae. *Nat. Biotechnol.*, **15**: 1359-1367.

Wu, L.Y., Thompson, D.K., Li, G., Hurt, R.A., Tiedje, J.M. and Zhou, J. (2001). Development and evaluation of functional gene arrays for detection of selected genes in the environment. *Appl. Environ. Microbiol.*, **67**: 5780-5790.

Ye, R.W., Tao, W., Bedzyk, L. et al. (2000). Global gene expression profiles of Bacillus subtilis grown under anaerobic conditions. *J. Bacterial*, **182**: 4458-4465.

Yeates, C., Holmes, A.J. and Gillings, M.R. (2000). Novel forms of ring-hydroxylating dioxygenases are widespread in pristine and contaminated soils. *Environ. Microbiol.*, **2**: 644-653.

Zhang, C. and Bennett, G.N. (2005). Biodegradation of xenobiotics by anaerobic bacteria. *Appl Microbiol Biotechnol*, **67**: 600-618.

Zhou, J. (2003). Microarrays for bacterial detection and microbial community analysis. *Curr. Opin. Microbiol.*, **6**: 288-294.

Zhou, J.-Z. and Thompson, D.K. (2002). Challenges in applying microarrays to environmental studies. *Curr. Opin. Biotechnol.*, **13**: 202-204.

11 Nanotechnology—In Relation to Bioinformatics

M.H. Fulekar

INTRODUCTION

The first use of word "Nanotechnology" has been attributed to Norio Taniguchi in a paper published in 1974. Eric Drexler, in 1986, published book "Engines of Creation" in which he described his ideas of molecular nanotechnology used to build miniature machines and devices from the bottom up using self-assembly. Many scientists from mainstream disciplines—biology, chemistry or physics—will argue of course that they are and have been 'doing' nanotechnology for years and that it is nothing new. Indeed chemists play with atoms and molecules which are sub-nano and molecular biology deals with the understanding and application of biological nano-scale components. Nature has used nanotechnology and, in fact, it has taken millions of years to develop this by a process of evolution and natural selection. Nano-technology is an emerging research field which promises to have a wide range of interesting applications. Nanotechnology encompasses all technology that aims to create nanometre-scaled structures or is able to address or manipulate matter at the nanometre level.

National Science and Technology Council (US) (2000) has defined Nanotechnology as: "Research and Technology development at the atomic, molecular, or macromolecular levels in the length of approximately 1-100 nm range, to provide fundamental understanding of phenomena and materials at the nanoscale, and to create and use structures, devices and systems that have novel properties and functions because of their small size. The novel and differentiating properties and functions are developed at a critical length scale of matter typically under 100 nm. Nanotechnology research and development includes integration of nanoscale structure into larger material components, systems, and architectures. Within these larger scale assemblies, the control and construction of their structures and component devices remain at the nanometer scale."

DNA COMPUTING

DNA computing is just one of the many applied areas of nanotechnology published by Dan E. Linstedt in 2004. He described that DNA computing is the ability to drive computations, store data and retrieve data through the structure and use of DNA molecules. DNA computing requires understanding of the molecule. For example, molecules are made up of atoms; the atoms contain protons, neutrons and electrons. The molecule can be a combination of atoms such as water molecule (H_2O). The nanotechnology aspect involves the ability of man to alter and control the makeup of the atom within the molecule.

DNA computing takes strands of DNA with proteins, enzymes and program specific states in each molecule in the double helix strand. To begin with, the idea of DNA computing is similar to the action of DNA. One strand of DNA houses the 'RAM' or memory, the other strand is a backup (like Raid 0 + 1 on disk). The enzymes are the motors that copy, search, access, read/write the information into the DNA strands. When the DNA is put into an aqueous solution (like water), and the data is added, the data or information finds the appropriate DNA component to combine with and attaches itself. The data is usually in the form of a chemical solution with its own enzymes, providing motion or movement to the atoms. Once the atoms bind, they cannot be unbound without changing the environment. Changing the environment may mean making it "unfriendly" to the data; thus the enzymes uncouple the chemically bonded elements (data) and return it to its previous state.

DNA computing marks require understanding on the data sets and to generate "software" that programs specific information and knowledge of chemistry and biology. It is well-known that in general to search across one gram of DNA in an aqueous solution, it might take one to three seconds. The nano-structures constructed in experimental demonstrations consists of DNA crossover molecules that self assemble into large lattices that can execute computations as well as DNA molecules that reconfigure for possible use of motors.

DNA computing experiment proved that data can be stored, replicated, searched and retrieved from DNA structures. "DNA bases represented the information bits: ATCG (nucleotides) spaced every 0.35 nanometres along the DNA molecule, giving DNA a remarkable data density of nearly 18 mbits per inch". This provides hope for computing power. Each nucleotide can represent a bit. Not only does the bit type make differences, but the order or sequence as well. A "T" in a third position means something completely different than a "T" in the first position, leading to limitless possibilities for computation.

Furthermore, each of these nucleotides can be complemented by "S" and hybridized. In other words, they can produce double stranded DNA. For error correction this is very important. It gives the nano-computer a chance to correct what should be a comparable equivalent (copy) of the data. Such is the way of Raid 0+1 disk arrays. For example, if there are four values per atom, each atom taking 0.35 microns across, this would be impressive for storage sizes. If we want to think about modeling the information, we must consider the first two-dimensional mode (2D) atoms with electrons; different atoms tied together through valence bonding, providing a surface area to the molecule and a chemical make up. The three-dimensional (3D) model allows for multiple twists of the atomic layers. Dan E. Linstedt in his document cited: compression and encryption algorithms have already been developed, tested and used in DNA computing. Terabyte-sized storage has already been reached and furthermore, quantum level parallel operations have also already been created, used and proven successful.

There have been similar reports of success from governments and research laboratories all over the world.

NANOENGINEERING BIOINFORMATICS

Engineering bioinformatics as a meaningful paradigm to complement modern nanotechnology has been examined by Lyshenski et al. (2003). They consider Bioinformatics as a coherent abstraction in devising, prototyping, design, optimization and analysis of complex nanosystems. Nano and microscale biological systems exist in nature in enormous variety and sophistication. They are applying complex biological patterns in order to devise, analyze and examine distinct systems. One cannot blindly copy biosystems due to the fact that many complex phenomena and effects have not been comprehended and system architectures and functionalities have not been fully examined. Typical examples include unsolved problems to comprehend the simplest Eschericchia Coli (E. coli) and Salmonella typhimurium bacteria that integrate three dimensional biocircuitary, computing processing, networking nanobioelectronics, nanobiomotors, nanobiosensors, etc. Correspondingly, attention is also concentrated on devising novel paradigms in systematic synthesis through Bioinformatics with the ultimate objective to fabricate these systems applying nanotechnology. This will allow one to derive new operating principles examining functionality of different subsystems, researching novel structures, studying advanced architectures (topologies) and characterizing distinct systems, subsystems and devices reaching the nanoarchitectionomics horizon. The scientist examines complex patterns in biosystems because superior system can be devised and designed through engineering Bioinformatics to achieve ultimate objective of engineering bioinformatics and system design. These are far-reaching frontiers of modern nanoscience and nanoengineering. The synergetic paradigm reported is demonstrated researching biosystems and coherently examining distinct nanostructures, complex and subsystems (Lyshenski et al., 2003).

NANOTECHNOLOGY - CANCER TREATMENT

Nanotechnology will complement genomic and proteomic research and accelerate the ability of scientist to prevent, detect and treat cancer. The onset of premalignant and malignant transformation will be detected at the molecular level long before the anatomic presence of a tumour is *discernible*. This will permit less drastic methods of elimination. A greater understanding of the human biology of cancer revealed from patients will complement and stimulate new laboratory investigations in vitro and in silico, linking delivery back to discovery in the continuum. Tomorrow's patients will know their susceptibility to cancer and the lifestyle factors needed to keep healthy. Molecular epidemiology and gene/environment interactions will help determine populations at risk. Emphasis on cancer biology will permit a systems approach to understand the cancer process. An integrated clinical trials system with a common bioinformatics grid will rapidly test and validate new strategies for early detection, prevention and prediction of cancer (Andrew & Eschenbach, 2004). The development of tools in genomics, proteomics, molecular imaging, bioinformatics, nanotechnology and other advanced technologies is a critical step.

BIOINFORMATICS, NANOTECHNOLOGY - DISINFECTANT FOR SEVERE ACUTE RESPIRATORY SYNDROME (SARS)

WHO, on March 12, 2003, issued a global alert on the outbreak of the epidemic – a new form of pneumonia-like disease with symptoms that are similar to those of the common flu. This illness is potentially fatal and highly contagious and had spread quickly to many parts of the world. Nevertheless, Severe Acute Respiratory Syndrome (SARS) was at one point a public health issue threatening large populations of the world.

Government research centres in Canada and the U.S. decoded the genome of the oeronavirus which was proved to be the cause of SARS. The British Columbia Cancer Agency (BCCA) in Vancouver was the first to sequence the SARS genome on April 13, 2003, followed closely by the Center for Disease Control and prevention (CDC) of the US on April 14, 2003.

The sequence information itself does not provide a cure, but rather the test and diagnostics for this particular virus. The sequencing success was a combination of several events, serendipity being one of the most significant. The challenge was to produce a DNA copy of the virus's RNA genome to work with. After several days of effort, scientists managed to produce one millionth of a gram of the genetic material on April 6, 2003. To sequence the SARS genome, the genome was broken in manageable fragments. Within a week, all the fragments had been sequenced. Once started, the sequencing itself was "fairly routine". The sequenced genome fragments were then assembled into the complete genome in 12 mers under UV light, electromer pairs are created. The negative electrons and positive holes create very strong oxidizers, called hydroxide radical, even stronger than chlorine used as a sterilizer. When harmful substances stick to the positive holes, they are completely broken down into the carbon dioxide and other harmless compounds. As a disinfectant, the hydroxide radical can also inhibit the growth of bacteria.

Bacteria can be found all over the place and they multiply quickly. The idea is to have disinfectant agents, such as TiO_2 that will kill bacteria faster than they multiply to sustain cleanliness. For TiO_2 to be effective as a disinfectant, the size has to be in the nanometre (10^{-9} m) range. In this range, it has been shown that the effectiveness of TiO_2 as disinfectant can go as high as 70 - 99%. The problem that scientists have is that the cost to grind the substance increases with diminishing size. Many industries now use micrometre (10^{-6} m) range TiO_2. Though much cheaper, the effect is drastically reduced.

There have been technologies developed along this line to deliver one of these ingredients at an extremely low concentration to create a powerful hospital grade disinfectant that is non-hazardous and environmentally safe (Lim, 2003). One particular product line, employing unique nano-emulsive technology, is reported to be able to reduce the spread of a broad range of diseases caused by microorganisms including E. coli, Salmonella, Listeria, Staph, Strep, Pseudomonas, MRSA, VRE, Norwonk-like Virus, Influenza A, Hepatitis B and C.

Another product has been developed using proprietary technology to create a nanoemulsion. The nanoemulsion can be sprayed, smeared on clothing, vehicles, people or anything that has been exposed to slew of deadly substances. It can also be rubbed on the skin, eaten or put into beverages like orange juice, and used in the water of a hot tub. The working principle is that the nano-bubbles contain energy that is stored as surface tension. The energy is released when bubbles coalesce, thus zapping the contaminant. The hurdle is that a huge amount of energy is needed to make the nanoemulsion, with bubbles of sizes smaller than bacteria and viruses.

The major concern is that opportunists might seize the scope arising from the market for cheap prevention kits, disinfectant substances, and sterilizing systems that are of dubious effectiveness. The risk is that the public may lower their guard under the false impression that they are fully protected. Note that all the products, if they are effective, are good only for preventing, disinfecting or detecting infectious agents; they do not offer cure, yet.

Nanomedicine

Nanomedicine has been described as the process of diagnosing, treating and preventing disease and traumatic injury, relieving pain and preserving and improving human health, using molecular tools and molecular knowledge of the human body. In short, nanomedicine is the application of nanotechnology to medicine (NSTC, 2000).

Nanomedicine is a large area and includes nanoparticles that act as biological mimetics, 'nanomachines', nanofibres and polymeric nanoconstructs as biomaterials, and nanoscale fabrication-based devices, sensors and laboratory diagnostics. Research into the delivery and targeting of pharmaceuticals, therapeutic and diagnostic agents using nanosized particles is the forefront of nanomedicine (NSTC, 2000). Freitas (2005) gave a partial taxonomy of nanomedicine technologies which was broken down into 18 classes (Table 1) and 96 sub-classes. This shows the potential application of nanotechnology in medicine.

Table 1: Nanomedicine Technologies – Main Classes

Raw nanomaterials	DNA manipulation, sequencing, diagnostics	Molecular medicine
Nanostructured materials	Tools and diagnostics	Artificial enzymes and enzyme control
Artificial binding sites	Intracellular devices	Nanotherapeutics
Control of surfaces	Bio MEMS	Synthetic biology and early nanodevices
Nanopores	Biological research	Biotechnology and biorobotics
Cell simulations and cell diagnostics	Drug delivery	Nanorobotics

NANOTECHNOLOGY IN DRUG DELIVERY

Nanotechnologists generate great excitement in the field of drug development and drug delivery. In fact, many are already used to treat patients. Indeed, since 1990, many drugs referred to as nano-therapeutics have been approved as products for clinical use. Most are anticancer drugs i.e. liposomes (e.g. Daunoxome), polymer coated liposomes (Doxil, Caclyx), polymeric drugs (copaxone), antibodies (Heraptin, Avastin) and antibody conjugates (mylotars), polymer protein conjugates (Oncaspar, Neulasta) and largely, a nanoparticles containing paclitaxel (Abraxane). These are nano-scale and commonly multicomponent constructs and can be viewed as the first nanomedicines to be used as clinical benefits. These nanomedicines are not really new. Many of these concepts e.g. antibody conjugates, liposomes, nanoparticles and polymer-conjugates were developed in the 1970s. However, significant funding has been given over to the research and development of Nanotechnologies that have great potential to contribute to medicine in the short and longer time. Today nanotechnology and nanoscience approaches to particle design and formulations are beginning to expand the market for many drugs and are forming the basis for a very profitable niche within the pharma industry, wherein some predicted benefits are hyped (Moghimi et al., 2005).

However, examples in the scientific literature which demonstrate the potential and the reason for excitement surrounding nanomedicine are overwhelming. The reader is referred to the reviews (Freitas, 2005; Moghimi et al., 2005).

Today, nanomedicine has branched out in hundreds of different directions, but the core concept remains that ability to structure materials and devices at the molecular scale can bring enormous immediate benefits in the research and practice of medicine. It has been proposed that miniaturization of medical tools will provide more accurate, more controllable, more versatile, more reliable, cheaper and faster approaches to enhancing the quality of human life.

REFERENCES

Andrew, C. and van Eschenbach (2004). Living with Cancer. *OECD Observer*, June 2004.

Drexler, K.E. (1986). Engines of Creation : The coming era of Nanotechnology. Anchor Books. Garden City, N.Y.

Freitas, R.A. (2005). Nanomedicine. 2-9.

http://www.cs.duke.edu/~reif/BMC/reports/BMC.FY99.reports/BMC.DARPA. FY99.html.

Lim, Hwa A (2003). Bioinformatics, Nanotechnology and SARS. 2003

Linstedt, Dan E. (2004). DNA Computing. http://arstechnica.com/reviews/2q00/dna/ dna-1.html.

Lyshenski, S.E., Kruegel, F.A. and Theodorou, E. (2003). Nanotechnology Bioinformatics: Nanotechnology paradigm and its applications. Third IEEE conference on Nanotechnology, 2003. **2:** 896-899.

Moghimi, S.M., Hunter, C.H. and Murray, J.C. (2005). Nanomedicine: Current status and future prospects. *FASEB Journal*, **19(3):** 311-330.

NSTC (2000). Definition of Nanotechnology: http://www.becon.nih.gov/ nstc_def_nano.htm, accessed 12 January, 2006.

12 Bioinformatics—Research Applications

S. Gupta

INTRODUCTION

Bioinformatics is a rapidly developing branch of biology which derives knowledge from computer analysis of biological data and is highly interdisciplinary, using techniques and concepts from informatics, statistics, mathematics, physics, chemistry, biochemistry, and linguistics. It has many practical applications in different areas of biology and medicine and it describes the use of computers to handle biological information. It is synonymous with "computational molecular biology" which means use of computers for the characterization of the molecular components of living things and analyzing the information stored in the genetic code as well as experimental results from various sources, patient statistics, and scientific literature. Research in bioinformatics includes storage, retrieval, and analysis of the data. Richard Durbin, Head of Informatics at the Wellcome Trust Sanger Institute, believes that all biological computing is not bioinformatics, e.g. mathematical modeling is not bioinformatics, even when connected with biology-related problems. Bioinformatics is mainly management and subsequent use of biological information, particularly, genetic information. Fredj Tekaia from the Institut Pasteur defines "Classical" bioinformatics as the mathematical, statistical and computing methods that aim to solve biological problems using DNA and amino acid sequences and related information. Medical imaging/image analysis and biologically-inspired computation as well as genetic algorithms and neural networks too are considered as part of bioinformatics. These areas interact in strange ways. Neural networks are inspired by crude models of the functioning of nerve cells in the brain and are used in a program called PHD to accurately predict the secondary structures of proteins from their primary sequences. Bioinformatics is thus the processing of large amounts of biologically-derived information pertaining to DNA sequences or X-rays.

The NIH Biomedical Information Science and Technology Initiative Consortium agreed on the following definitions of bioinformatics and computational biology, recognizing that no definition could completely eliminate overlap with other activities or preclude variations in interpretation by different individuals and organizations.

Bioinformatics: Research, development or application of computational tools and approaches for expanding the use of biological, medical, behavioural or health data, including those to acquire, store, organize, archive, analyze or visualize such data.

Computational Biology: The development and application of data-analytical and theoretical methods, mathematical modeling and computational simulation techniques to the study of biological, behavioural, and social systems.

The National Center for Biotechnology Information defines bioinformatics as "the field of science in which biology, computer science, and information technology merge into a single discipline." There are three important sub-disciplines within bioinformatics:

(i) the development of new algorithms and statistics with which to assess relationships among members of large data sets;

(ii) the analysis and interpretation of various types of data including nucleotide and amino acid sequences, protein domains and protein structures; and

(iii) the development and implementation of tools that enable efficient access and management of different types of information.

APPLICATIONS OF BIOINFORMATICS

Bioinformatics has many applications in the research areas of medicine, biotechnology, agriculture, etc.

1. **Genomics:** It is an attempt to analyze or compare the entire genetic complement of a species. It is possible to compare genomes by comparing representative subsets of genes within genomes.

2. **Proteomics:** Proteomics is the study of proteins—their location, structure and function. It is the identification, characterization and quantification of all proteins involved in a particular pathway, organelle, cell, tissue, organ or organism that can provide accurate and comprehensive data about that system. It deals with the study of the proteome, called proteomics viz. all the proteins in any given cell as well as the set of all protein isoforms and modifications, the interactions between them, the structural description of proteins and their higher-order complexes and everything 'post-genomic'.

3. **Pharmacogenomics:** Pharmacogenomics is the application of genomic approaches and technologies to the identification of drug targets. It uses genetic information to predict whether a drug will help make a patient

well or sick and study how genes influence the response of humans to drugs i.e. pharmacogenetics.

4. **Pharmacogenetics:** Pharmacogenetics is the study of how the actions of and reactions to drugs vary with the patient's genes. All individuals respond differently to drug treatments; some positively, others with little obvious change in their conditions and yet others with side effects or allergic reactions. Much of this variation is known to have a genetic basis. Pharmacogenetics is a subset of pharmacogenomics which uses genomic/bioinformatic methods to identify genomic correlates, for example SNPs (Single Nucleotide Polymorphisms), characteristic of particular patient response profiles and use those markers to inform the administration and development and improvement of therapies.

5. **Cheminformatics:** It deals with the mixing of information resources and appropriate analysis to transform data into information and information into knowledge for the specific purpose of drug lead identification and optimization. Related terms of cheminformatics are chemometrics, computational chemistry, chemical informatics and chemical information management/science.

   Chemical informatics: Computer-assisted storage, retrieval and analysis of chemical information, from data to chemical knowledge. (*Chem. Inf. Lett.,* 2003, **6:** 14)

   Chemometrics: The application of statistics to the analysis of chemical data (from organic, analytical or medicinal chemistry) and design of chemical experiments and simulations. [IUPAC Computational]

6. **Structural genomics or structural bioinformatics** refers to the analysis of macromolecular structure, particularly proteins using computational tools and theoretical frameworks. One of the goals of structural genomics is the extension of the idea of genomics to obtain accurate three-dimensional structural models for all known protein families, protein domains or protein folds. Structural alignment is a tool of structural genomics.

7. **Comparative genomics:** The study of human genetics by comparisons with model organisms such as mice, the fruit fly and the bacterium E. coli.

8. **Biophysics:** The British Biophysical Society defines biophysics as "an interdisciplinary field which applies techniques from the physical sciences to understanding biological structure and function".

9. **Biomedical informatics/Medical informatics:** Biomedical informatics is an emerging discipline that has been defined as the study, invention, and implementation of structures and algorithms to improve communication, understanding and management of medical information.

10. **Mathematical Biology:** Mathematical biology also tackles biological problems. The methods it uses to tackle them need not be numerical

and need not be implemented in software or hardware. It includes things of theoretical interest which are not necessarily algorithmic, or molecular in nature nor useful in analyzing collected data.

11. **Computational chemistry:** Computational chemistry is the branch of theoretical chemistry whose major goals are to create efficient computer programs that calculate the properties of molecules (such as total energy, dipole moment, vibrational frequencies) and to apply these programs to concrete chemical objects. It is also sometimes used to cover the areas of overlap between computer science and chemistry. It is a discipline using mathematical methods for the calculation of molecular properties or for the simulation of molecular behaviour. It also includes synthesis planning, database searching, combinatorial library manipulation (Hopfinger, 1981; Ugi et al., 1990). [IUPAC Computational]

12. **Functional genomics:** Functional genomics is a field of molecular biology using the vast wealth of data produced by genome sequencing projects to describe genome function. It uses high-throughput techniques like DNA microarrays, proteomics, metabolomics and mutation analysis to describe the function and interactions of genes.

13. **Pharmacoinformatics:** Pharmacoinformatics concentrates on the aspects of bioinformatics dealing with drug discovery.

14. **In silico ADME-Tox Prediction:** Drug discovery is a complex and risky treasure hunt to find the most efficacious molecule which do not have toxic effects but at the same time have desired pharmacokinetic profile. Huge amount of research is required to be done to come out with a molecule which has the reliable binding profile. The molecule which shows better binding is then evaluated for its toxicity and pharmacokinetic profiles so that the molecule becomes a successful drug.

15. **Agroinformatics/Agricultural informatics:** Agro informatics concentrates on the aspects of bioinformatics dealing with plant genomes.

16. **Systems biology:** Systems biology is the coordinated study of biological systems by investigating the components of cellular networks and their interactions, by applying experimental high-throughput and whole-genome techniques and integrating computational methods with experimental efforts.

BIOINFORMATICS—TOOLS, SOFTWARES AND PROGRAMS

Bioinformatic tools are software programs that are designed for extracting the meaningful information from the mass of molecular biology/biological databases and to carry out sequence or structural analysis.

The following factors must be taken into consideration when designing bioinformatics tools, software and programs:

(a) The end user (the biologist) may not be a frequent user of computer technology.
(b) These software tools must be made available over the internet given the global distribution of the scientific research community.

Major Categories of Bioinformatic Tools

There are both standard and customized products to meet the requirements of particular projects. There are data-mining software that retrieve data from genomic sequence databases and also visualization tools to analyze and retrieve information from proteomic databases. These can be classified as homology and similarity tools, protein functional analysis tools, sequence analysis tools and miscellaneous tools. Everyday bioinformatics is done with sequence search programs like BLAST, sequence analysis programs like the EMBOSS and Staden packages, structure prediction programs like THREADER or PHD or molecular imaging/modeling programs like RasMol and WHATIF.

Homology and Similarity Tools

Homologous sequences are sequences related by divergence from a common ancestor. Thus the degree of similarity between two sequences can be measured while their homology is a case of being either true or false. This set of tools can be used to identify similarities between novel query sequences of unknown structure and function and database sequences whose structure and function have been elucidated.

Protein Function Analysis

This group of programs allow you to compare your protein sequence to the secondary (or derived) protein databases that contain information on motifs, signatures and protein domains. Highly significant hits against these different pattern databases allow you to approximate the biochemical function of your query protein.

Structural Analysis

This set of tools allows the comparison of structures with the known structure databases. The function of a protein is more directly a consequence of its structure rather than its sequence with structural homologs tending to share functions. The determination of a protein's 2D/3D structure is crucial in the study of its function.

Sequence Analysis

This set of tools allows you to carry out more detailed analysis on your query sequence including evolutionary analysis, identification of mutations, hydropathy regions, CpG islands and compositional biases. The identification of these and other biological properties are all clues that aid the search to elucidate the specific function of your sequence.

Some Examples of Bioinformatic Tools

BLAST: BLAST (**B**asic **L**ocal **A**lignment **S**earch **T**ool) comes under the category of homology and similarity tools. It is a set of search programs designed for the Windows platform and is used to perform fast similarity searches regardless of whether the query is for protein or DNA. Comparison of nucleotide sequences in a database can be performed. Also a protein database can be searched to find a match against the queried protein sequence. NCBI has also introduced the new queuing system to BLAST (Q BLAST) that allows users to retrieve results at their convenience and format their results multiple times with different formating options. Depending on the type of sequences to compare, there are different programs:

- blastp compares an amino acid query sequence against a protein sequence database.
- blastn compares a nucleotide query sequence against a nucleotide sequence database.
- blastx compares a nucleotide query sequence translated in all reading frames against a protein sequence database.
- tblastn compares a protein query sequence against a nucleotide sequence database dynamically translated in all reading frames.
- tblastx compares the six-frame translations of a nucleotide query sequence against the six-frame translations of a nucleotide sequence database.

FASTA: It is an alignment program for protein sequences created by Pearsin and Lipman (1988), and is one of the many heuristic algorithms proposed to speed up sequence comparison. The basic idea is to add a fast prescreen step to locate the highly matching segments between two sequences, and then extend these matching segments to local alignments using more rigorous algorithms such as Smith-Waterman.

EMBOSS: EMBOSS (**E**uropean **M**olecular **B**iology **O**pen **S**oftware **S**uite) is a software-analysis package. It can work with data in a range of formats and also retrieve sequence data transparently from the Web. Extensive libraries are also provided with this package, allowing other scientists to release their software as open source. It provides a set of sequence-analysis programs, and also supports all UNIX platforms.

Clustalw: It is a fully automated sequence alignment tool for DNA and protein sequences. It returns the best match over a total length of input sequences, be it a protein or a nucleic acid.

RasMol: It is a powerful research tool to display the structure of DNA, proteins, and smaller molecules. Protein Explorer, a derivative of RasMol, is an easier to use program.

PROSPECT: PROSPECT (PROtein Structure Prediction and Evaluation Computer ToolKit) is a protein-structure prediction system that employs a computational technique called protein threading to construct a protein's 3-D model.

PatternHunter: PatternHunter, based on Java, can identify all approximate repeats in a complete genome in a short time using little memory on a desktop computer. Its features are its advanced patented algorithm and data structures, and the java language used to create it. The Java language version of PatternHunter is just 40 kB, only 1% the size of Blast, while offering a large portion of its functionality.

COPIA: COPIA (COnsensus Pattern Identification and Analysis) is a protein structure analysis tool for discovering motifs (conserved regions) in a family of protein sequences. Such motifs can be then used to determine membership to the family for new protein sequences, predict secondary and tertiary structure and function of proteins and study evolution history of the sequences.

Application of Programs in Bioinformatics

JAVA in Bioinformatics

Since research centres are scattered all around the globe ranging from private to academic settings, and a range of hardware and OSs are being used, Java is emerging as a key player in bioinformatics. Physiome Sciences' computer-based biological simulation technologies and Bioinformatics Solutions' PatternHunter are two examples of the growing adoption of Java in bioinformatics.

Perl in Bioinformatics

String manipulation, regular expression matching, file parsing, data format interconversion etc. are the common text-processing tasks performed in bioinformatics. Perl excels in such tasks and is being used by many developers. Yet, there are no standard modules designed in Perl specifically for the field of bioinformatics. However, developers have designed several of their own individual modules for the purpose, which have become quite popular and are coordinated by the BioPerl project.

Bioinformatic Projects

An exhaustive list of various bioinformatic projects is given here; a detailed description of these projects is available on the internet. This list includes BioPerl, BioXML, Biocorba, Ensembl, Bioperl-db, Biopython and BioJava.

Biopython and bioJava are open source projects with very similar goals to bioPerl. However their code is implemented in python and java, respectively. With the development of interface objects and biocorba, it is possible to write java or python objects which can be accessed by a bioPerl script, or to call bioPerl objects from java or python code. Since biopython and bioJava are more recent projects than bioPerl, most effort to date has been to port bioPerl functionality to biopython and bioJava rather than the other way around. However, in the future, some bioinformatic tasks may prove to be more effectively implemented in java or python in which case being able to call them from within bioPerl will become more important.

Major bioinformatic activities are carried out using three types of Molecular Modeling programs viz., AMBER, CHARMM and GROMACS as well as three types of Genetic Algorithms STRUCTURE OPTIMIZATION, SEQUENCE ALIGNMENT and PSEGA.

Appendix 1

BIOINFORMATICS—WEBSITES

A System for Easy Analysis of Lots of Sequences (SEALS)
http://www.ncbi.nlm.nih.gov/CBBresearch/Walker/SEALS/index.html

Applied Bioinformatics Journal
http://www.openmindjournals.com/bioinformatics.html

Australian Genome Research Facility (AGRF)
http://www.agrf.org.au/

Beijing Genomics Institute (BGI)
http://coe.genomics.org.cn/

Bielefeld University Bioinformatics Server
http://bibiserv.techfak.uni-bielefeld.de/

BIMAS - Bioinformatics and Molecular Analysis Section
http://bimas.cit.nih.gov/

BioInform Bioinformatics News Service
http://www.bioinform.com/index.htm

Bioinformatics & Computational Biology at ISU
http://www.bcb.iastate.edu/

Bioinformatics & Computational Biology Index of Resources
http://www.unl.edu/stc-95/ResTools/biotools/biotools4.html

Bioinformatics (Genomics) Selected Papers
http://post.queensu.ca/~forsdyke/bioinfor.htm

Bioinformatics 2001, 2002
http://www.ida.his.se/ida/bioinformatics2001/

Bioinformatics and XML
http://www.biosino.org/bioinformatics/011212-2.htm

Bioinformatics at NIH
http://grants1.nih.gov/grants/bistic/bistic.cfm

Bioinformatics Canada
http://www.bioinformatics.ca/

Bioinformatics Centre
http://bioinfo.ernet.in/

Bioinformatics Centre (BIC)
http://www.bic.nus.edu.sg/

Bioinformatics Education and Research
http://bioinf.man.ac.uk/

Bioinformatics Forum
http://www.biowww.net/forum/list.php?f=4

Bioinformatics Glossary
http://www.genomicglossaries.com/content/Bioinformatics_gloss.asp
http://www.incyte.com/glossary/index.shtml

Bioinformatics Group
http://life.anu.edu.au/

Bioinformatics Guides
http://his.cbi.pku.edu.cn:1947/binf/Guides/

Bioinformatics in India
http://www.bioinformatics-india.com/index.php3

Bioinformatics Information
http://www.bioinformatics.pe.kr/

Bioinformatics Institute (BII)
http://www.bii-sg.org/

Bioinformatics Links
http://www.ii.uib.no/~inge/list.html

Bioinformatics Links and Sites
http://www.colorado.edu/chemistry/bioinfo/

Bioinformatics Organization
http://bioinformatics.org/

Bioinformatics Resources
http://www.genet.sickkids.on.ca/bioinfo_resources/

Bioinformatics Services
http://www.bioinformaticsservices.com/

Bioinformatics: Biology and Molecular Biology Resources
http://www.nwfsc.noaa.gov/protocols/bioinformatics.html

Bioinformatics: Human Genome Project Information
http://www.ornl.gov/hgmis/research/informatics.html

Bioinformatics: Sequence and Genome Analysis
http://www.bioinformaticsonline.org/

BioInformer Newsletter
http://bioinformer.ebi.ac.uk/

BioTech: Bioinformatics
http://biotech.icmb.utexas.edu/pages/bioinfo.html

BMC Bioinformatics Research Papers
http://www.biomedcentral.com/bmcbioinformatics/

Boston University Bioinformatics
http://bioinfo.bu.edu/

Buffalo Center of Excellence in Bioinformatics
http://www.bioinformatics.buffalo.edu/

CBI - Center of BioInformatics, Peking University
http://www.cbi.pku.edu.cn/

Center for Bioinformatics & Computational Biology at Duke
http://bioinformatics.duke.edu/

Center for Biological Sequence (CBS) Analysis
http://www.cbs.dtu.dk/

Comprehensive Bioinformatics Solutions: What's new in Bioinformatics?
http://www.accelrys.com/bio/

Computational Biology At ORNL
http://compbio.ornl.gov/

Computational Proteomics and Protein Structural Information
http://www.strubix.com/

Developing Bioinformatics Computer Skills (Book) 2001
http://www.oreilly.com/catalog/bioskills/

European Bioinformatics Institute (EBI)
http://www.ebi.ac.uk/

ExPASy Molecular Biology Server
http://www.expasy.org/

Gene Alliance - The Key to Genomic Knowledge
http://www.gene-alliance.com/

GeneBio Geneva Bioinformatics
http://www.genebio.com/

GeneCards - A Database of Human Genes
http://genecards.bjmu.edu.cn/

Genome Megasequencing Program
http://megasun.bch.umontreal.ca/ogmpproj.html

Genome Technology Magazine
http://www.genome-technology.com/

Google Web Directory: Bioinformatics
http://directory.google.com/Top/Science/Biology/Bioinformatics/

How to become a Bioinformatics Expert
http://www.techfak.uni-bielefeld.de/bcd/ForAll/Econom/study.html

IBM Bioinformatics & Pattern Discovery Group
http://researchweb.watson.ibm.com/bioinformatics/

IEEE Computer Society Bioinformatics Conference
http://conferences.computer.org/bioinformatics/

INCOGEN: Institute for Computational Genomics
http://www.incogen.com/

Institute of Bioinformatics, Tsinghua University
http://www.bioinfo.tsinghua.edu.cn/

Integrated Bioinformatics
http://www.biosino.org/bioinformatics/010807-3.htm

International Conference on Bioinformatics 2002 (INCoB 2002)
http://incob.biotec.or.th/

ISREC Bioinformatics Group
http://www.isrec.isb-sib.ch/

Lab for Advanced Sequence Analysis
http://shark.ucsf.edu/gc/

MCW Bioinformatics Research Center (BRC)
http://brc.mcw.edu/

Molecular Biology and Bioinformatics Resources
http://www.molbio.org/

Molecular Genetics Server
http://www.pasteur.fr/recherche/borrelia/Welcome.html

National Bioinformatics Institute (NBI)
http://www.bioinfoinstitute.com/

National Center for Biotechnology Information (NCBI)
http://www.ncbi.nlm.nih.gov/About/primer/bioinformatics.html

O'Reilly Bioinformatics Conference 2003
http://conferences.oreillynet.com/bio2003/

Online Journal of Bioinformatics (OJB)
http://www.cpb.ouhsc.edu/ojvr/bioinfo.htm

Online Lectures on Bioinformatics
http://lectures.molgen.mpg.de/

Open Bioinformatics Foundation
http://www.open-bio.org/

Open Directory: Bioinformatics
http://dmoz.org/Science/Biology/Bioinformatics/

Oxford Journals Online: Bioinformatics
http://bioinformatics.oupjournals.org/

PEDANT: Protein Extraction, Description, and Analysis Tool
http://pedant.gsf.de/

Peking-Yale Joint Research Center For Plant Molecular Genetics and Agro-Biotechnology
http://www.pyc.pku.edu.cn/

Program in Proteomics and Bioinformatics
http://p-b.med.utoronto.ca/

Protein Data Bank (PDB)
http://www.rcsb.org/pdb/

Protein Sequence Analysis: A Practical Guide
http://www.biochem.ucl.ac.uk/bsm/dbbrowser/jj/prefacefrm.html

Society for Bioinformatics
http://www.socbin.org/

South African National Bioinformatics Institute (SANBI)
http://www.sanbi.ac.za/

Southampton Bioinformatics Data Server (SBDS)
http://molbiol.soton.ac.uk/

Stockholm Bioinformatics Center (SBC)
http://www.sbc.su.se/

Swiss Institute of Bioinformatics (SIB)
http://www.isb-sib.ch/

TBI—Theoretical Bioinformatics
http://www.dkfz-heidelberg.de/tbi/

TBR—The Bioinformatics Resources
http://www.hgmp.mrc.ac.uk/CCP11/index.jsp

Texas A&M Bioinformatics Working Group
http://www.csdl.tamu.edu/FLORA/tamuherb.htm

The Asia Pacific Bioinformatics Network (APBioNET)
http://www.apbionet.org/

The BioPerl Project
http://www.bioperl.org/

The Brutlag Bioinformatics Group
http://motif.stanford.edu/

The Genome Database
http://gdbwww.gdb.org/

The Virtual Institute of Bioinformatics
http://www.bioinf.org/vibe/index.html

TUBIC - Tianjin University BioInformatics Center
http://tubic.tju.edu.cn/

UCLA Bioinformatics
http://www.bioinformatics.ucla.edu/

UCSC Genome Bioinformatics
http://genome.ucsc.edu/

Upenn Center for Bioinformatics
http://www.pcbi.upenn.edu/

Virginia Bioinformatics Institute (VBI)
http://www.vbi.vt.edu/

Visual Bioinformatics
http://www.visual-bioinformatics.com/

Visualization For Bioinformatics
http://industry.ebi.ac.uk/~alan/VisSupp/

Weizmann Institute of Science: Genome and Bioinformatics
http://bioinfo.weizmann.ac.il/bioinfo.html

What Is Bioinformatics?
http://bioinformatics.weizmann.ac.il/cards/bioinfo_intro.html

Yale Bioinformatics
http://bioinfo.mbb.yale.edu/

Appendix 2

BIOIFORMATICS – TERMINOLOGY

A

Accession number: An identifier supplied by the curators of the major biological databases upon submission of a novel entry that uniquely identifies that sequence (or other) entry.

Active site: The amino acid residues at the catalytic site of an enzyme. These residues provide the binding and activation energy needed to place the substrate into its transition state and bridge the energy barrier of the reaction undergoing catalysis.

Additive genetic effects: The combined effects of alleles at different loci are equal to the sum of their individual effects.

Adenine: A nitrogenous base, one member as the base pair AT (adenine–thymine).

Agents: Software modules (independent, autonomous) that can search the internet for data or content pertinent to a particular application such as a gene, protein, or biological system.

Aggregation techniques: A technique used in model organisation studies in which embryos at the 8-cell stage of development are pushed together to yield a single embryo.

Algorithm: A fixed procedure embodied in a computer program.

Alignment: The process of lining up two or more sequences to achieve maximal levels of identity (and conservation, in case of amino acid sequencing) for the purpose of assessing the degree of similarity and the possibility of homology.

Allele frequency: Relative proportion of a particular allele among individuals of a population.

Alleles: One of two or more alternate forms of a gene.

Allogenic: Variation in alleles among members of the same species.

Alternative splicing: Alternate combinations of a folded protein that is possible due to recombination of multiple gene segments during mRNA splicing that occurs in higher organisms.

Amino acid: The basic building block of proteins.

Amplification: An increase in the number of copies of a specific DNA fragment.

Annotation: Adding pertinent information such as gene coded for amino acid sequence, or other commentary to the database entry of raw sequence of DNA bases.

Anticodon: The triplet of contiguous bases on tRNA that binds to the codon sequence of nucleotides, on mRNA. Example: GGG codes for glycine.

Antigen: Any foreign molecule that stimulates an immune response in a vertebrate organism. Many antigens act as the surface proteins of foreign organisms.

Antisense: Nucleic acid that has a sequence exactly opposite to an mRNA molecule made by the body; binds to the mRNA molecule to prevent a protein from being made.

Apoptosis: A programed cell death, the body's normal method of disposing of damaged, unwanted or unneeded cells.

Assay: A method for measuring a biological activity. This may be enzyme activity, binding affinity, or protein turnover. Most assays utilize a measurable parameter such as colour, fluorescence or radioactivity to correlate with the biological activity.

Assembly: Putting sequenced fragments of DNA into their correct chromosomal positions.

Autosome: When the chromosome is not involved in sex determination. The diploid human genome consists of a total of 46 chromosomes, 22 pairs of autosomes, and one pair of sex chromosomes (the X and Y chromosomes).

B

Bacterial artificial chromosome: Cloning vector that can incorporate large fragments of DNA.

Bacteriophage: A virus that infects bacteria. The bacteriophage DNA has served as a basis for cloning vectors, and is also utilized to create phase libraries containing human or other genes.

Baculovirus: Virus which forms the basis of a protein expression system.

Base pair: A pair of nitrogenous bases (a purine and pyrimidine), held together by hydrogen bonds, that form the core of DNA and RNA i.e. the A:T, G:C and A:U interactions.

Base sequence: The order of nucleotide bases in a DNA molecule; determines structure of proteins encoded by the DNA.

Base sequence analysis: A method, sometimes automated for determining the base sequence.

Bioinformatics: The science of managing and analyzing biological data using advanced computing techniques.

Bivalent: Two binding sites, having two free electrons available for binding.

BLAST: A computer program that identifies homologous (similar) genes in different organisms such as human, fruit fly or nematode.

C

Cancer: A disease in which abnormal cells divide and grow unchecked. Cancer can spread from its original site to other parts of the body and can be fatal.

Candidate gene: A gene located in a chromosome region suspected of being involved in a disease.

Capillary array: Gel-filled silica capillaries used to separate fragments for DNA sequencing. The small diameter of the capillaries permit the application of higher electric fields, providing high speed, high throughput separations that are significantly faster than traditional slab gels.

Carrier: An individual who possesses an unexpressed, recessive trait.

cDNA library: A collection of DNA sequences that code for genes. The sequences are generated in the laboratory from mRNA sequences.

Cell: The basic unit of any living organism that carries on the biochemical processes of life.

Centromere: A specialized chromosome region to which spindle fibres attach during cell division.

Chloroplast chromosome: Circular DNA found in the photosynthesizing organelle (chloroplast) of plants instead of the cell nucleus where most genetic material is located.

Chromatin: Chromosome as it appears in its condensed state composed of DNA and associated proteins.

Chromosomal deletion: The loss of part of a chromosome's DNA.

Chromosomal inversion: Chromosome segments that have been turned 180 degrees. The gene sequence for the segment is reversed with respect to the rest of the chromosome.

Chromosome: The structure in the cell nucleus that contains the entire cellular DNA together with a number of proteins that compact and package the DNA.

Chromosome painting: Attachment of certain fluorescent dyes to targeted parts of the chromosome, which is used for particular disease diagnosis.

Clone: An exact copy mode of biological material such as DNA segment (e.g. a gene or other region), a whole cell, or a complete organism.

Cloning: The formation of clones or exact genetic replicas.

Codon: A sequence of three adjacent nucleotides that designates a specific amino acid or start/stop site transcription.

Coding regions: Portion of a genomic sequences bounded by start and stop codons that identifies the sequence of the protein being coded for by a particular gene.

Comparative genomics: The study of human genetics by comparisons with model organisms such as mica, fruit fly and the bacterium (E. coli).

Complementary DNA (cDNA): DNA strand copied from mRNA using reverse transcriptase. A cDNA library represents all of the expressed DNA in a cell.

Complementary sequence: Nucleic acid base sequence that can form a double-stranded structure with another DNA fragment by following base-pairing rules (A pairs with T and C with G). The complementary sequence to GTAC, for example, is CATG.

Configuration: The complete ordering and description of all parts of a software or database system. Configuration management is the use of software to identify inventory and maintain the component modules that together comprise one or more systems or products.

Confirmation: The precise three dimensional arrangement of atoms and bonds in a nucleotide describing its geometry and hence its molecular function.

Conservation: Changes at a specific position of an amino acid or (less commonly, DNA) sequence that preserve the physico-chemical properties of the original residue.

Conserved sequence: A base sequence in a DNA molecule (amino acid sequence in a protein) that has remained essentially unchanged throughout evolution.

Constitutive synthesis: Synthesis of mRNA and protein at an unchanging or constant rate regardless of a cell's requirements.

Cosmids: DNA vectors that allow the insertion of long fragments of DNA (up to 50 K bases).

Crystal structure: Describe the high resolution molecular structure derived by X-ray crystallographic analysis of protein or other biomolecualr crystals.

Cytogenetics: The study of the physical appearance of chromosomes.

Cytological map: A type of chromosome map whereby genes are located on the basis of cytological findings obtained with the aid of chromosome mutations.

Cytoplasm: The medium of the cell between the nucleus and the cell membrane.

Cytosine: A pyrimidine base found in DNA and RNA.

D

Data mining: The ability to query very large databases in order to satisfy a hypothesis; or to interrogate a database in order to generate new hypothesis based on rigorous statistical correlations.

Data processing: The systematic performance of operations upon data such as handling, merging, sorting and computing. The semantic content of the original data should not be changed, but the semantic content of the processed data may be changed.

Data warehouse: A collection of databases, data tables and mechanisms to access the data on a single subject.

Database: Any file system by which data gets stored following a logical process.

Deconvolution: Mathematical procedure to separate the overlapping effects of molecules such as mixtures of compounds in a high-throughput screen, or mixtures of cDNAs in a high density array.

Deletion: A loss of part of the DNA from a chromosome can lead to a disease of abnormality.

Deletion map: A description of a specific chromosome that uses defined mutations; specific deleted areas in the genome as 'biochemical signposts', or markers for specific areas.

Dimer: A composite molecule formed by the binding of two molecules.

Directed evolution: A laboratory process used on isolated molecules or microbes to cause mutations and identify subsequent adaptations to novel environments.

Directed mutagenesis: Alteration of DNA at a specific site and its reinsertion into an organism to study any effects of the change.

Directed sequencing: Alleles carrying particular DNA sequences associated with the presence of disease.

DNA: The molecule that encodes genetic information. DNA is a double-stranded molecule held together by weak hydrogen bonds base pairs of nucleotides. The four nucleotides in DNA contain the bases adenine (A), guanine (G), cytosine (C) and thymine (T). In nature, base pairs form only between A and T and between G and C; thus the base sequence of each single strand can be deduced from that of its partner.

DNA bank: A service that stores DNA extracted from blood samples or other human tissue.

DNA finger printing: A technique for identifying human individuals based on a restriction enzyme digest of tandemly repeated DNA sequences that are scattered throughout the human genome, but are unique to each individual.

DNA microarrays: The deposition of oligonucleotides or cDNAs onto an inert substrate such as glass or silicon. Thousands of molecules may be organised spatially into a high density matrix. These DNA chips may be probed to allow expression monitoring of many thousand of genes simultaneously. Uses include study of polymorphisms in genes, de novo sequencing or molecular diagnosis of disease.

DNA polymerase: Enzyme that catalyses the synthesis of DNA from a DNA template given the deoxy ribo-nucleotide precursors.

DNA probes: Short single stranded DNA molecules of specific base sequence, labeled either radioactively or immunologically, that are used to detect and identify the complementary base sequence in a gene or genome by hybridizing specifically to that gene or sequence.

DNA repair genes: Genes encoding proteins that correct errors in DNA sequencing.

DNA replication: The uses of existing DNA as a template for the synthesis of new DNA strands. In humans and other eukaryotes, replication occurs in the cell nucleus.

DNase: One of a series of enzymes that can digest DNA.

Domain: A discrete portion of a protein with its own function. The combination of domains in a single protein determines its overall function.

Dominant: Allele that is almost always expressed, even if only one copy is present.

Drug: An agent that affects a biological process, specifically, a molecule whose molecular structure can be correlated with its pharmacological activity.

Drug discovery cycle: The cycle of events requested to develop a new drug. Typically, this involves research, pre-clinical testing and clinical development, and can take from 5 to 12 years.

E

E. Colia (Escherichia Coli): A bacterium that has been studied intensively by geneticists because of its small genome size, normal lack of pathogenicity and easy growth in the laboratory.

Electrophoresis: A method of separating large molecules (DNA fragments or proteins) from a mixture of similar molecules. An electric current is passed through a medium containing the mixture, and each kind of molecule travels through the medium at a different rate, depending on its electrical charge and size. Agarose and acrylamide gels are the media commonly used for electrophoresis of proteins and nucleic acids.

Electroporation: A process using high voltage current to make cell membranes permeable to allow the introduction of new DNA, commonly used in recombinant DNA technology.

Embryonic stem cells: An embryonic cell that can replicate indefinitely, transform into other type of cells and serve as a continuous source of new cells.

Enhancers: DNA sequences that can greatly increase the transcription rates of genes even though they may be far upstream or downstream from the promoter they stimulate.

Eukaryote: Cell or organism with membrane-bound, structuring discrete nucleus and other well-developed subcellular compartments. For example: Eukaryotes include all organisms except viruses, bacteria and blue green algae.

Enzyme: A protein that acts as a catalyst, spreading the rate at which a biochemical reaction proceeds but not altering the direction or nature of the reaction.

EST (Expressed Sequence Tag): A short strand of DNA that is a part of a cDNA molecule and can act as identifier of a gene. Used in locating and mapping genes.

Exogenous DNA: DNA originating outside an organism that has been introduced into the organism.

Exon: The protein-coding DNA sequence of a gene.

Exonuclease: Enzyme that cleans nucleotides frequently from free ends of a linear nucleic acid substrate.

Expression profile: The level and duration of expression of one or more genes, selected from a particular cell or tissue type, generally obtained by a variety of high-throughput methods, such as sample sequencing, serial analysis, or micro array-based detection.

Expression vector: A cloning vector that is engineered to allow the expression of protein from a cDNA. The expression vector provides an appropriate promoter and restriction sites that allow insertion of cDNA.

F

FASTA: The first widely used algorithm for database similarity searching. The program looks for optimal local alignments by scanning the sequence for small matches called "words". The sensitivity and speed of the search are inversely related and controlled by the "K-tup" variable which specifies the size of a "word".

Filtering: Also known as masking. The process of hiding regions of (nucleic acid or amino acid) sequence having characteristics that frequently lead to spurious high scores.

Finger printing: In genetics, the identification of multiple specific alleles on a person's DNA to produce a unique identifier for that person.

Flow karyotyping: Use of flow cytometry to analyze and separate chromosomes according to their DNA content.

Fluorescence In-Situ Hybridization (FISH): A physical mapping approach that uses fluorescein tags to detect hybridization of probes with metaphase chromosomes and with the less-condensed somatic interphase chromatin.

Frame shift: A deletion, substitution, or duplication of one or more bases that causes the reading frame of a structural gene to shift the normal series of triplets.

Full gene sequence: The complete order of bases in gene. This order determines which protein a gene will produce.

Functional genomics: The study of genes, their resulting proteins, and the role played by the proteins, the body's biochemical processes.

Fusion protein: The protein resulting from the genetic joining and expression of two different genes.

G

Gamete: Mature male or female reproductive cell (sperm or ovum) with a haploid set of chromosomes (23 for humans).

Gel electrophoresis: A technique by which molecules are separated by size or charge by passing them through a gel under the influence of an external electric field.

Gene: The fundamental physical and functional unit of heredity. A gene is an ordered sequence of nucleotides located in a particular position on a particular chromosome that encodes a specific functional product (i.e. a protein or RNA molecule).

Gene amplification: Repeated copying of a piece of DNA; a characteristic of tumour cells.

Gene bank: Data bank of genetic sequence operated by a division of the National Institute of Health.

Gene chip technology: Development of cDNA micro arrays from a large number of genes. Used to monitor and measure changes in gene expression for each gene represented on the chip.

Gene expression: The conversion of information from gene to protein via transcription and translation.

Gene families: Subsets of genes containing homologous sequences which usually correlate with a common function.

Gene library: A collection of cloned DNA fragments created by restriction endonuclease digestion that represent part or all of an organism's genome.

Gene pool: All the variations of genes in a species.

Gene prediction: Predictions of possible genes made by a computer program based on how well a stretch of DNA sequence matches known gene sequences.

Gene product: The product, either RNA or protein, that results from expression of a gene. The amount of gene product reflects the activity of the gene.

Gene project: Research and technology development effort aimed at mapping and sequencing the genome of human beings and certain model organisms.

Gene therapy: The use of genetic material for therapeutic purposes. The therapeutic gene is typically delivered using recombinant virus or liposome based delivery systems.

Gene transfer: Incorporation of new DNA into an organism's cells, usually by a vector such as modified virus. Used in gene therapy.

Genetic: The study of inheritance patterns of specific traits.

Genetic code: The mapping of all possible codons into the 20 amino acids including the start and stop codons.

Genetic counselling: Provides patients and their families with education and information about genetic-related conditions and helps them make informed decisions.

Genetic discrimination: Prejudice against those who have or are likely to develop an inherited disorder.

Genetic engineering: The procedure used to isolate, splice and manipulate DNA outside the cell. Genetic engineering allows a recombinant engineered DNA segment to be introduced into a foreign cell or organism and be able to replicate and function normally.

Genetic illness: Sickness, physical disability or other disorders resulting from the inheritance of one or more deleterious alleles.

Genetic marker: Any gene that can be readily recognized by its pheno ypic effect, and which can be used as a marker for a cell, chromosome or individual carrying that gene.

Genetic polymorphism: Difference in DNA sequence among individuals, groups or populations (e.g. genes for blue eyes versus brown eyes).

Genetic predisposition: Susceptibility to a genetic disease. May or may not result in actual development of the disease.

Genetic testing: Analyzing an individual's genetic material to determine predisposition to a particular health condition or to confirm a diagnosis of genetic disease.

Genome: The complete genetic content of an organism.

Genomic DNA: DNA sequence typically obtained from mammalian or other higher-order species, which include intron and exon sequence (coding sequence) as well as non-coding regulatory sequences such as promoter and enhanced sequences.

Genomics: The analysis of the entire genome of a chosen organism.

Genotype: The total set of genes present in the cells of an organism. Also, refers to a set of alleles at a single gene locus.

Germ cell: Sperm and egg cells and their precursors. Germ cells are haploid and have only one set of chromosomes (23 in all), while all other cells have two copies (46 in all).

Germ line: The continuation of a set of genetic information from one generation to the next.

Global alignment: The alignment of the nucleic acid or protein sequences over their entire length.

Glycosylation: The addition of carbohydrate groups (sugars) e.g. to polypeptide chains.

Guanine (G): One of the nitrogenous purine bases found in DNA and RNA

Gyandromorph: Organism that have both male and female cells and therefore express both male and female characteristics.

H

Hairpin: A double-vertical region in a single DNA or RNA strand formed by the hydrogen-bonding between adjacent inverse complementary sequences to form a hairpin shaped structure.

Haploid: A single set of chromosomes present in the egg and sperm cells of animals and in the egg and pollen cells of plants.

Haplotype: A way of denoting the collective genotype of a number of closely linked loci on a chromosome.

Hemizygous: Having only one copy of particular gene, e.g. human males are hemizygous as only one copy of genes found on Y-chromosomes.

Hereditary cancer: Cancer that occurs due to the inheritance of an altered gene within a family.

Heterodimer: Protein composed of two different chains or subunits.

Heteroduplex: Hybrid structure formed by the annealing of two DNA strands (or an RNA and DNA) that have sufficient complementary sequence to allow hydrogen bonding.

Highly conserved sequence: DNA sequence that is very similar across several different types of organisms.

High-throughput sequencing: A fast method of determining the order of bases in DNA.

Homeodomain: A 60-amino acid protein domain coded by the homeobox region of a homeotic gene.

Homeotic gene: A gene that controls the activity of other genes involved in the development of a body plan. Homeotic genes have been found in organisms ranging from plants to humans.

Homologous chromosome: Chromosome containing the same linear gene sequences as another, each derived from one parent.

Homologous recombination: Swapping of DNA fragments between paired chromosomes.

Homologue: A member of a chromosome pair in diploid organisms or a gene that has the same origin and functions in two or more species.

Homology: Similarity in DNA or protein sequences between individuals of the same species or among different species.

Homozygote: An organism that has two identical alleles of a gene.

Housekeeping genes: Genes that are always expressed due to their constant requirement by the cell.

HSP: High-Scoring Segment Pair. Local alignments with no gaps that achieve one of the top alignment scores in a given search.

Human artificial chromosome (HAC): A vector used to hold large DNA fragments.

Hybrid: The offspring of genetically different parents.

Hybridization: The process of joining two complementary strands of DNA or one each of DNA and RNA to form a double-stranded molecule.

I

Identical twin: Twins produced by the division of a single zygote; both have identical genotype.

Identity: The extent to which two (nucleotide or amino acid) sequences are invariant.

Immunoglobulin: A member of the globulin protein family consisting of the light and too heavy chains linked by disulfide bends. All antibodies are immunoglobulin.

Immunotherapy: Using the immune system to treat disease, e.g. in the development of vaccines. May also refer to the therapy of diseases caused by the immune system.

Inherit: In genetics, to receive genetic material from parents through biological processes.

Insertion: A mutation that occurs when an extra nucleotide is inserted into a gene sequence, causing a frame shift.

In-situ hybridization: A variation of the DNA/RNA hybridization procedure in which the denatured DNA is in place in the cell and is then challenged with RNA or DNA extracted from another source.

Integration: The physical insertion of DNA into the host cell genome. The process is used by retroviruses where a specific enzyme catalyses the process or can occur at random sites with other DNA.

Interphase: The period in the cell cycle when DNA is replicated in the nucleus, followed by mitosis.

Introns: Nucleotide sequences found in the structural genes of eukaryotes that are non-coding and interrupt the sequences containing information that code for polypeptide chains. Intron sequences are spliced out of their RNA transcripts before maturation and protein synthesis.

In-vitro: Studies performed outside a living organism such as in a laboratory.

In-vivo: Studies carried out in a living organism.

Isoschizomers: Two different restriction enzymes which recognize and cut DNA at the same recognition site.

Isozymes: Two or more enzymes capable of catalyzing the same reaction but varying in their specificity due to differences in their structures and hence their efficiencies under different environmental conditions.

J

Junk DNA: Term used to describe the excess DNA that is present in the genome beyond that required to encode proteins. A misleading term since these regions are likely to be involved in gene regulation, and other as yet unidentified functions.

K

Karyotype: The constitution of chromosomes in a cell or individual.

Kilobase (Kb): Unit of length for DNA fragments equal to 1000 nucleotides.

Knockernt: Deactivation of specific genes; uses in laboratory organisms to study gene function.

L

Lambda: A statistical parameter used in calculating BLAST scores that can be thought of as a natural scale for scoring. The value lambda is used in converting a raw score (S) to a bit score (S′).

Lead compounds: A candidate compound identified as the best "hit" (tight binder) after screening of a combinational (or others) compound library, that is then taken into further rounds of screening to determine its suitability as a drug.

Leucine zipper: Protein motif which binds DNA in which 4-5 leucines are found at 7 amino acid intervals. This motif is present typically in transcription factors and other proteins that bind DNA.

Lexicon: In bioinformatics, a lexicon refers to a pre-defined list of terms that together completely define the contents of a particular database.

Library: An unordered collection of clones (i.e. clones DNA from a particular organism) whose relationship to each other can be established by physical mapping.

Ligand: Any small molecule that binds to a protein or receptor; the cognate partner of many cellular proteins, enzymes and receptors.

Linkage: The association of genes (or genetic loci) on the same chromosome. Genes that are linked together tend to be transmitted together.

Linkage map: A map of the relative positions of genetic loci on a chromosome, determined on the basis of how often the loci are inherited together. Distance is measured in centimorgans (cM).

Local alignment: The alignment of some portion of two nucleic acid or protein sequences.

Localize: Determination of the original position (locus) of a gene or other marker on a chromosome.

Locus: The position on a chromosome of a gene or other chromosome marker; also, the DNN at that position. The use of locus is sometimes restricted to mean expressed DNA regions.

Low Complexity Region (LCR): Regions of biased composition including homopolymeric runs, short-period repeats, and more subtle overrepresentation of one or a few residues. The program is used to mask or filter LCRs in amino acid queries.

M

Macro restriction map: Map depicting the order of and distance between sites at which restriction enzymes cleave chromosomes.

Map unit: A measure of genetic distance between two linked genes that corresponds to a recombination frequency of 1%.

Mapping population: The group of related organisms used in constructing a genetic map.

Markov chain: Any multivariate probability density, whose independence diagram is a chain. The variables are ordered and each variable "depends" only on its neighbours in the sense of being conditionally independent of the others. Markov chains are an integral component of hidden Markov models.

Masking: Also known as filtering. The removal of repeated or low complexity regions from a sequence in order to improve the sensitivity of sequence similarity searches performed with that sequence.

Mass spectrometry: An instrument used to identify chemicals in a substance by their mass and charge.

Megabase (Mb): Unit of length for DNA fragments equal to one million nucleotides and roughly equal to 1 cM.

Meiosis: A process within the cell nucleus that results in the reduction of the chromosome number from diploid (two copies of each chromosome) to haploid (a single copy) through two reductive divisions in germ cells.

Melting (of DNA): The denaturation of double-stranded DNA into two single strands by the application of heat (denaturation breaks the hydrogen bonds holding the double-stranded DNA together).

Merbid map: A diagram showing the chromosomal location of genes associated with disease.

Messenger RNA (mRNA): Any single-stranded RNA molecule that encodes the information necessary to synthesize a given protein.

Metaphase: A stage in mitosis or meiosis during which the chromosomes are aligned along the equatorial plane of the cell.

Methylation: The addition of $-CH_3$ (methyl) groups to a target site. Typically such addition occurs on the cytosine bases of DNA.

Mitosis: The nuclear division that results in the replication of the genetic material and its redistribution into each of the daughter cells during cell division.

Microarray: Sets of miniaturized chemical reaction areas that may also be used to test DNA fragments, antibodies or proteins.

Microbial Genetics: The study of genes and gene function in bacteria, archaea and other microorganisms often used in research in the fields of bioremediation, alternative energy and disease prevention.

Micronuclei: Chromosomal fragments that are not incorporated into the nucleus at cell division.

Microfumidics: The miniaturization of chemical reactions or pharmacological assays into microscopic tubes of vessels in order to greatly increase their throughput, in placing many of them side-by-side in an array.

Microinjection: A technique for introducing a solution of DNA into a cell using a fine microcapillary pipette.

Mimetics: Compounds that mimic the function of other molecules via their high degree of structural (conformational) similarity, and hence physico-chemical properties.

Missense Mutation: A point mutation in which one codon (triplet of bases) is changed into another designating a different amino acid.

Mitochondria: Known as the cell's power sources, this distinct organelle have two membranes and usually a rod-shape, but also can have round shape.

Mitochondrial DNA: The genetic material found in mitochondria, the organelles that generate energy for the cell. Not inherited in the same fashion as nucleic DNA.

Model organism: A laboratory animal or other organism useful for research.

Modeling: The use of statistical analysis, computer analysis or model organisms to predict outcomes of research.

Molecular biology: The study of the structure, function and make-up of biologically important molecules.

Molecular farming: The development of transgenic animals to produce human proteins for medical use.

Molecular genetics: The study of macro molecules important in biological inheritance.

Molecular medicine: The treatment of injury or disease at the molecular level. Examples include the use of DNA-based diagnostic tests or medicine derived from DNA sequence information.

Monogenic disorder: A disorder caused by mutation of a single gene.

Monomer: A single unit of any biological molecule or macromolecule, such as an amino acid, nucleic acid, polypeptide domain, or protein.

Monosomy: Possessing only one copy of a particular chromosome instead of the normal two copies.

Monovalent: Having one binding site; strictly an atom with only one pair of electron available for binding in its highest energy shell.

Motif: A short-conserved region in a protein sequence. Motifs are frequently highly conserved parts of domains.

Multigene family: A set of genes derived by duplication of an ancestral gene, followed by independent mutational events resulting in a series of independent genes either clustered together on a chromosome or dispersed throughout the genome.

Multiple sequence alignment: An alignment of three or more sequences with gaps inserted in the sequences such that residues with common structural positions and/or ancestral residues are aligned in the same column. Crystal W is one of the most widely used multiple sequence alignment programs.

Multiple sequencing: Approach to high-throughput sequencing that uses several pooled DNA samples run through gels simultaneously and then separated and analyzed

Multiplexing: A laboratory approach that performs multiple sets of reactions in parallel; greatly increasing speed and throughput.

Mutagen: Any agent that can cause an increase in the rate of mutations in an organism.

Mutagenecity: The capacity of a chemical or physical agent to cause permanent genetic alterations.

Mutation: An inheritable alteration to the genome that includes genetic (joint or single case) changes, or larger scale alterations such as chromosomal deletions or rearrangement.

N

Naked DNA: Pure, isolated DNA devoid of any proteins that may bind to it.

NCEs (New Chemical Entity): Compounds identified as potential drugs that are sent from research and developments into clinical trials to determine their suitability.

Nested PCR: The second round amplification of an already PCR-amplified sequence using a new pair of primers which are internal to the original primers. Typically done when a single PCR reaction generates insufficient amounts of product.

Nitrogenous base: An important part of DNA that makes up the genetic sequence. The bases are adenine, guanine, thymine (uracil in RNA) and cytosine.

Nonsense strand: A complementary RNA strand to the DNA strand that is transcribed and contains no promoter site.

Northern blot: A gel-based laboratory procedure that locates mRNA sequence on a gel that are complementary to a piece of DNA used as a probe.

Nuclease: Any enzyme that can clean the phosphodiester bonds of nucleic acid backbones.

Nucleic acid: A large molecule composed of nucleotide subunits.

Nucleolus: A structure within the nucleus that contains a bunch of protein and RNA gathered together into ribosomes.

Nucleoral organizing region: A part of the chromosome containing rRNA genes.

Nucleoside: A fine-carbon sugar covalently attached to a nitrogen base.

Nucleotide: A single molecule composed of a phosphate, a fine carbon sugar, and a nitrogenous base that makes up the sequences of DNA.

Nucleus: The brain of a cell, where the DNA is located.

O

Oligogenic: A phenotypic trait produced by two or more genes working together.

Oligonucleotide: A short molecule consisting of several linked nucleotides (typically between 10 and 60) covalently attached by phosphodiester bonds.

Oncogene: A gene, one or more forms of which are associated with cancer; many oncogenes are involved, directly or indirectly.

Open Reading Frame (ORF): The sequence of DNA or RNA located between the start-code sequence (initiation codon) and the stop-code sequence (termination codon).

Operator: A segment of DNA that interacts with the products of regulatory genes and facilitates the transcription of one or more structural genes.

Operon: A unit of transcription consisting of one or more structural genes, an operator, and a promoter.

Optimal alignment: An alignment of the sequences with the highest possible score.

Orthologous: Homologous sequences in different species that arose from a common ancestral gene during speciation; may or may not be responsible for a similar function.

Overlapping clones: Collection of cloned sequence made by generating randomly overlapping DNA fragment with infrequently cutting restriction enzymes.

P

Paralog: Paralogs are genes related by duplication within a genome; orthologs retain the same function in the course of evolution, whereas paralogs evolve new functions even if these are related to the original one.

Peptide: A short stretch of amino acids each covalently linked by a peptide bond (–CO–NH) or amide bond.

Peptide bond (Amide bond): A covalent bond formed between two amino acids when the amino group of one is linked to the carboxy group of another (resulting in the elimination of one water molecule).

Phage (Bacteriophage): A virus that infects bacterial cells and serves as a useful vector for introducing genes into bacteria for a number of purposes.

Phenotype: The realized expression of the genotype, the observable expression of a trait which results from the biological activity of proteins or RNA molecules transcribed from the DNA.

Plasmid: Any replicating DNA element that can exist in the cell independently of the chromosomes. Synthetic plasmids are used for DNA cloning. Most commonly found in bacterial cells.

Point mutation: A mutation that occurs when one base pair along a chain of DNA is changed, usually in an exon, or region coding for a protein. There are three types of point mutation: substitution, deletion and insertion.

Polypeptide: A chain of amino acid that is connected by links called a peptide bond.

Post-transcriptional modification: Alterations made of pre-mRNA before it leaves the nucleus and becomes mature mRNA.

Post-translational modification: Alteration made to a protein after its synthesis at the ribosome. These modifications, such as the addition of carbohydrate or fatty acid chains, may be critical to the function of the protein.

Primary sequence (Protein): The linear sequence of a polypeptide or protein.

Primer: A short-nucleic acid sequence that is the initiation point for the addition of nucleotides in DNA replication.

Probe: Any biochemical that is labeled or tagged in some way so that it can be used to identify or isolate a gene, RNA or protein.

Prokaryote: An organism or cell that lacks a membrane bounded nucleus. Bacteria and blue-green algae are the only surviving prokaryotes.

Promoter: A portion of DNA where RNA polymerase attaches itself to begin transcription.

Promoter site: The site on DNA where the transcription process begins. Composed of nitrogenous bases.

Protein: Protein made up of a chain of amino acid, and has a specific 3D shape that defines a job.

Proteome: The entire protein complement of a given organism.

Proteomics: Systematic analysis of protein expression of normal and diseased tissues that involves the separation, identification and characterization of all of the proteins in an organism.

Purine: A nitrogen-containing compound with a double-ring structure. The parent compound of adenine and guanine.

Pyrimidine: A nitrogen-containing compound with a single six-membered ring structure. The parent compound of thymidine and cytosine.

Q

Query: The input sequence (or other type of search term) with which all of the entities in a database are to be compared.

Query sequence: A DNA, RNA of protein sequence used to search a sequence database in order to identify close or remote family members (homologs) of known function, or sequences with similar active sites or regions (analogs) from whom the function of the query may be deduced.

R

Rational drug design: The development of drugs based on the 3-dimensional molecular structure of a particular target.

Reading frame: A sequence of codons beginning with an initiation codon and ending with a termination codon, typically of at least 150 bases (50 amino acids) coding for a polypeptide or protein chain.

Recessive: Any trait that is expressed phenotypically only when present on both alleles of a gene.

Recessive gene: A gene which will be expressed only if there are two identical copies, or for a male, if one copy is present on the X-chromosome.

Recombinant clone: Clone containing recombinant DNA molecules.

Recombinant DNA (rDNA): DNA molecules resulting from the fusion of DNA from different sources. The technology employed for splicing DNA from different sources and for amplifying the resultant heterogeneous DNA.

Recombination: The formation of new gene combinations not originally found in the cell.

Regulatory gene: A DNA sequence that functions to control the expression of other genes by producing a protein that modulates the synthesis of their products.

Relational Database Management System (RDBMS): A software system that includes a database architecture, query language and data loading and updating tools and other ancillary software that together allow the creation of a relational database application.

Replication: The synthesis of an informationally identical macromolecule (e.g. DNA) from a template molecule.

Repressor: The protein product of a regulatory gene that combines with a specific operator (regulatory DNA sequence) and hence blocks the transcription of genes in an operon.

Restriction enzyme (Restriction endonuclease): A type of enzyme that recognizes specific DNA sequence (usually palindromic sequences 4, 6, 8 or 16 base pairs in length) and produces cuts on both strands of DNA containing those sequences only. The "molecular scissors" of rDNA technology.

Restriction Fragment Length Polymorphisms (RFLPs): Variation within the DNA sequences of organisms of a given species that can be identified by fragmenting the sequences using restriction enzymes, since the variation lies within the restriction site. RFLPs can be used to measure the diversity of a gene in a population.

Restriction map: A physical map or depiction of a gene (a genome) derived by overlapping restriction fragments produced by digestion of the DNA with a member of restriction enzymes.

Reverse genetics: The use of protein information to elucidate the genetic sequence encoding that protein. Used to describe the process of gene isolation starting with a panel of afflicted patients.

Reverse Transcriptase-PCR (RT-PCR): Procedure in which PCR amplification is carried out on DNA that is first generated by the conversion of mRNA to cDNA using reserve transcriptase.

Reverse transcriptase: A DNA polymerase that can synthesize a complementary DNA (cDNA) strand using RNA as a template also called RNA dependent DNA polymerase.

Ribonucleic Acid (RNA): A category of nucleic acids in which the component sugar is ribose and consisting of four nucleotides: thymidine, uracil, guanine and adenine. The three types of RNA are messenger RNA (mRNA), transfer RNA (tRNA) and ribosomal RNA (rRNA).

Ribosomal RNA (rRNA): RNA on globular form, a major component of ribosomes. Its role is protein synthesis.

Ribosome: An organelle located within a cell that has two subunits; its function is protein synthesis.

Risk communication: In genetics, a process in which a genetic counsellor or other medical professional interprets genetic test results and advises patients of the consequences for them and their offspring.

RNA Polymerase: An enzyme used to create mRNA from either a strand of DNA or RNA.

S

Sanger sequencing: A widely used method of determining the order of bases in DNA.

Satellite: A chromosomal segment that branches off from the rest of the chromosome but is still connected by a thin filament, a stack.

Scaffold: In genome mapping, a series of configs that are in the right order but not necessarily connected in one continuous stretch of sequence.

SEG: A program for filtering low complexity regions in amino acid sequences. Residues that have been masked are pre-presented as "X" in an alignment.

Segregation: The normal biological process whereby the two pieces of a chromosome pair are separated during meiosis and randomly distributed to the germ cells.

Sense strand: The strand of DNA where the promoter site is located and mRNA receives its messages from the promoter site.

Sequence assembly: A process whereby the order of multiple sequential DNA fragments is determined.

Sequence Tagged Site (STS): A unique sequence from a known chromosome location that can be amplified by PCR. STSs act as physical markers for genomic mapping and cloning.

Sequencing: Determination of the order of nucleotides (base sequences) in a DNA or RNA molecule or the order of amino acids in a protein.

Sequencing technology: The instrumentation and procedures used to determine the order of nucleotides in DNA.

Sex chromosome: The X or Y chromosomes in human being that determines the sex of an individual. Females have two X chromosomes in diploid cells, males have one X and one Y chromosome. The sex chromosomes comprise the 23rd chromosome pair in a Karyotype.

Sex-linked: Traits or diseases associated with X or Y chromosomes; generally seen in males.

Shotgun cloning: The cloning of an entire gene segment or genome by generating a random set of fragments using restriction endonucleases to create a gene library that can be subsequently mapped and sequenced to reconstruct the entire genome.

Single gene disorder: Hereditary disorder caused by a mutant allele of a single gene.

Single Nucleotide Polymorphism (SMP): DNA sequence variations that occur when a single nucleotide (A, T, C or G) in the genome sequence is altered.

Somatic cell: Any cell in the body except gametes and their precursors.

Somatic cell gene therapy: Incorporating new genetic material into cells for therapeutic purposes. The new genetic material cannot be passed to offspring.

Somatic cell genetic mutation: A change in the genetic structure that is neither inherited nor passed to offspring. Also called acquired mutations.

Southern blotting: Transfer by absorption of DNA fragments separated in electrophoretic gels to membrane filters for detection of specific base sequences by radio-labeled complementary probes.

Splice site: Location in the DNA sequence where RNA removes the non-coding areas to form a continuous gene transcript for translation into a protein.

Splicing: The cutting of introns and joining of exons to form a complete RNA strand with no introns.

Stem cell: Undifferentiated primitive cells in the bone marrow that have the ability, both to multiply and to differentiate into specific blood cells.

Structural gene: Gene which encodes a structured protein.

Structural genomics: The effort to determine the 3D structures of large numbers of proteins using both experimental techniques and computer simulation.

Substitution: The pressure of a non-identical amino acid at a given position in an alignment. If the aligned residues have similar physico-chemical properties, the substitution is said to be "conservative".

Suppressor gene: A gene that can suppress the action of another gene.

Syndrome: The group or recognizable pattern of symptoms or abnormalities that indicate a particular trait or disease.

Syngeneic: Genetically identical members of the same species.

Synteny: Genes occurring in the same order on chromosomes of different species.

T

Tandem repeat sequences: Multiple copies of the same base sequence on a chromosome; used as markers in physical mapping.

Targeted mutagenesis: Deliberate change in the genetic structure directed at a specific site on the chromosome. Used in research to determine the targeted regions function.

Technology transfer: The process of transferring scientific findings from research laboratories to the commercial sector.

Telomerase: The enzyme that directs the replication of telomeres.

Telomeres: The end of a chromosome. This specialized structure is involved in the replication and stability of linear DNA molecules.

Tertiary Structure: Folding of a protein chain via interactions of sidechain molecules including formation of disulfide bands between cystein residues.

Thymine: A pyrimidine base found in DBA but not in RNA.

Trans section: The introduction of foreign DNA into a host cell.

Transcript: The single-stranded mRNA chain that is assembled from a gene template.

Transcription: The process of a DNA strands being used as a template for the formation of RNA.

Transcription factors: A group of regulatory proteins required for transcription in eukaryotes. Transcription factors bind to the promoter region of a gene and facilitate transcription by RNA polymerase.

Transfer RNA (tRNA): A small RNA molecule that recognizes a specific amino acid, transports it to a specific codon in the mRNA and positions it properly in the nascent polypeptide chain.

Transformation: A genetic alteration to a cell as a result of the incorporation of DNA from a genetically different cell or virus; can also refer to the introduction of DNA into bacterial cells for genetic manipulation.

Transgene: A foreign gene that is introduced into a cell or whole organism for therapeutic or experimental purposes.

Translation: The process of mRNA being translated into amino acids, which group together in chain to form proteins; occur in ribosome.

Translocation: A mutation in which a large segment of one chromosome breaks off and attaches to another chromosome.

Transposable element: A class of DNA sequences that can move from one chromosomal site to another.

Truncated protein: Occurs when an amino acid is replaced with a certain other amino acid in a gene that causes the insertion of a stop codon, which therefore terminates the production of proteins from that point on.

U

Unidentified Reading Frame (URF): An open reading frame encoding a problem of undefined function.

Uracil: Nitrogenous pyrimidine base found in RNA but not DNA.

V

Vector: Any agent that transfers material (typically DNA) from one host to another.

Visualization: The process of representing abstract scientific data as images that can aid in understanding the meaning of the data.

W

Western blot: Technique used to identify and locate proteins based on their ability to bind to specific antibodies.

Wild type: Form of a gene or allele that is considered the "standard" or most common.

X

X chromosome: In mammals, the sex chromosome that is found in two copies in the homogametic sex (female in humans) and one copy in the heterogametic sex (male in humans).

Xenograft: Tissue or organ from an individual of one species transplanted into or grafted onto an organism of another species, genes or family. A common example of this is the use of heart valves in humans.

Y

Y chromosome: One of the two sex chromosomes, X and Y.

Yeast Artificial Chromosome (YAC): Constructed from yeast DNA, it is a vector used to clone large DNA fragments.

Z

Z-DNA: A conformation of DNA existing as a left handed double helix which may play a role in gene regulation.

Zinc finger protein: A secondary feature of some protein containing a zinc atom, DNA finding proteins.

Index